T0275790

LONDON MATHEMATICAL SOCIETY STUDENT TEXTS

Managing editor: Dr C.M. Series, Mathematics Institute
University of Warwick, Coventry CV4 7AL, United Kingdom

London Mathematical Society Student Texts 20

Communication Theory

Charles M. Goldie
School of Mathematical Sciences,
Queen Mary and Westfield College, University of London

and

Richard G. E. Pinch
Department of Pure Mathematics and Mathematical Statistics.
University of Cambridge

The right of the
University of Cambridge
to print and sell
all manner of books
was granted by
Henry VIII in 1534.
The University has printed
and published continuously
since 1584.

CAMBRIDGE UNIVERSITY PRESS
Cambridge
New York   Port Chester
Melbourne   Sydney

CAMBRIDGE
UNIVERSITY PRESS

University Printing House, Cambridge CB2 8BS, United Kingdom

Published in the United States of America by Cambridge University Press, New York

Cambridge University Press is part of the University of Cambridge.

It furthers the University's mission by disseminating knowledge in the pursuit of education, learning and research at the highest international levels of excellence.

www.cambridge.org
Information on this title: www.cambridge.org/9780521406062

First published 1991

A catalogue record for this publication is available from the British Library

ISBN 978-0-521-40456-3 Hardback
ISBN 978-0-521-40606-2 Paperback

To Jennifer
C.M.G.
and Geraldine
R.G.E.P.

CONTENTS

PREFACE

This text is based on a course of the same title given at Cambridge for a number of years. It consists of an introduction to information theory and to coding theory at a level appropriate to mathematics undergraduates in their second or later years. Prerequisites needed are a knowledge of discrete probability theory and no more than an acquaintance with continuous probability distributions (including the normal). What is needed in finite-field theory is developed in the course of the text, but some knowledge of group theory and vector spaces is taken for granted.

The two topics treated are traditionally put into mathematical pigeon-holes remote from each other. They do however fit well together in a course, in addressing from different standpoints the same problem, that of communication through noisy channels. The authors hope that undergraduates who have liked algebra courses, or probability courses, will enjoy the *other* half of the book also, and will feel at the end that their knowledge of how it all fits together is greater than the sum of its parts.

The Cambridge course was invented by Peter Whittle and the debt that particularly the information-theoretic part of the book owes him is unre-payable. Certain features that distinguish the present approach from that found elsewhere are due to him, in particular the conceptual 'decoupling' of source and channel, and the definition of channel capacity as a maximized rate of reliable transmission. The usual definition of channel capacity is, from that standpoint, an *evaluation*, less fundamental than the definition.

In detail, the first four chapters cover the information-theory part of the course. The first, on noiseless coding, also introduces entropy, for use throughout the text. Chapter 2 deals with information sources and gives a careful treatment of the evaluation of rate of information output. Chapters

3 and 4 deal with channels and random coding. An initial approach to the evaluation of channel capacity is taken in Chapter 3 that is not quite sharp, and so yields only bounds, but which seems considerably more direct and illuminating than the usual approach through mutual information. The latter route is taken in Chapter 4, where several channel capacities are exactly calculated.

Chapter 5 deals with error-control codes. The basic definitions are followed by an analysis of some constraints on the design of codes with specified error-control rates, with their asymptotic form. In Chapter 6 the properties of finite fields are developed and used to describe families of cyclic codes. A primer on finite fields and other algebraic objects is provided as an appendix, Chapter 7.

The book has a large number of exercises and problems, well provided with hints. Practice exercises are included at the ends of most sections, and longer problems — which should still be well within the reader's grasp — at the ends of chapters. Our debt to old Cambridge examiners and to the inventors of all the standard textbook exercises will be obvious there.

Formulae are numbered (1), (2), ... and exercises 1., 2., ... within each section, and referred to using those numbers. When referring to formulae and exercises across sections we expand the numbering, so that $(m.n.p)$ refers to formula (p) in §$m.n$, and Exercise $m.n.p$ to the pth exercise in §$m.n$. Figures, tables, theorems and so on are always given full numberings including the chapter and section numbers.

The book has been typeset by us, using the incomparable TEX typesetting language of D. E. Knuth. The master copy was produced on the photo-typesetting equipment at the University of London Computer Centre, with unstinting expert assistance from Philip Taylor of Royal Holloway and Bedford New College. The diagrams were done with M. J. Wichura's PJCTEX extension to TEX. The layout and type-style are based on a Cambridge University Press format which we have extended to cover our particular needs. To the usual admission of responsibility for faults of content we must thus add that for faults of typesetting.

Our debt to Peter Whittle has been mentioned and we also gratefully acknowledge his encouragement to us to write the book. David Tranah, Publishing Director for Mathematical Sciences at Cambridge University Press, commissioned the work, and we thank him most cordially for doing so and for waiting patiently for its completion.

C. M. G.   *April 1991*

R. G. E. P.

0

INTRODUCTION

Information theory is the study of how the amount of content in a stream of data may be evaluated, and how fast it may in principle be shipped from place to place by a given *communication channel*. The channel may need the data in a specific form, for instance as a stream of on-off pulses, and may corrupt it by randomly introducing errors. The subject is thus built on discrete probability theory as its mathematical base. It is somewhat top-down in approach, giving bounds and existence proofs without always any explicit means of implementation.

Coding theory studies explicitly how a stream of data may be transformed for transmission through a given channel, in ways that are efficient yet allow the stream to be reconstituted from a partly corrupted version. It is built on the algebraic theory of finite fields as its mathematical base. It is 'bottom-up' in approach, giving explicit recipes for action without necessarily claiming optimality.

The book studies both topics, as explained in the preface. Information theory starts from the transmission diagram

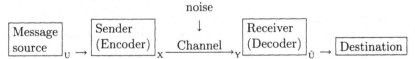

in which the main components of a communication system are displayed in the boxes. The subscripts to the boxes are notations we shall use for the messages at various stages of processing.

Of the components shown, the Message source as well as the Channel and the Destination are to be considered fixed, specified by whatever system is producing, transmitting and receiving the messages, and not subject to your control. You can think of the channel as a telephone line, that does indeed transmit the messages, but it could equally well be a compact disc, that instead stores them. Your control is exercised through the Sender, or Encoder, and the Receiver, or Decoder. The sender has three functions: to compress the message if possible, while not reducing its content in any way, to transform it into the form the channel may require (e.g. on-off pulses, modulation of a carrier waveform, ...), and to build in to it error-correction capability so that the original may be retrieved as accurately as possible despite channel-induced degradation.

We will study the first function in Chapter 1 under the name 'Economical representation'. Broadly, the idea is that particular patterns that turn up often in the source message-stream might as well be represented by short 'codewords'. For instance, if the source produces data in columns of fixed width separated by 12 spaces, you would certainly want to have a short codeword to represent a string of 12 consecutive spaces. We will model probabilistically the fact that particular patterns turn up with greater or lesser frequencies. Much of Chapter 1 will be spent on making precise the notion of quantity of information produced by a 'random' source-stream, in terms of the notion of (information) *entropy*. While this is distinct from physicists' entropy, there are links and similarities, justifying the name. We will also study particular compression schemes (Shannon-Fano and Huffman encoding).

In order to deal with the other functions of the sender, we have to spend Chapter 2 developing the idea of the *rate* at which a source produces information in the long run, given optimum encoding. There will turn out to be an advantage in taking the source-output in long blocks of symbols. We will attempt to quantify this advantage, as it has to balanced against the practical storage and delay problems of such blocking. A deeper treatment of entropy will be needed, and an investigation of its rate of production for particular sources: Bernoulli sources, Markov sources, stationary sources and ergodic sources.

Chapters 3 and 4, then, turn to channels and how fast data may be sent down them, given optimum encoding and decoding of the source-stream. We have split the channel discussion between Chapters 3 and 4 for reasons which you can read about in the introductory passages to those chapters. Optimal decoding, by the way, is essentially a *statistical* problem, that of inferring a message from a possibly corrupted version. We discuss it in that light in Chapter 3.

Chapters 5 and 6 discuss efficient methods for encoding messages which are able to detect and correct a proportion of errors. You are probably already familiar with the idea of a 'check-digit', as used, for example, in electronic data transmission. In Chapter 5 we give bounds on the rate at which such codes can transmit data with a given rate of error correction, and see that the notion of entropy turns up again. We shall pay particular attention to codes which can be given the structure of a vector space over a finite field. In Chapter 6 we deal with cyclic codes, which have a higher degree of symmetry, and find that they are inimately bound up with the algebra of finite fields.

Finally, in Chapter 7 we review the algebraic concepts used in this book. We assume that you already have some familarity with elementary abstract algebra and hope that you will find this Appendix helpful in reviewing your algebra and explaining our notations and terminology.

ECONOMICAL REPRESENTATIONS: NOISELESS CODING

The aim in this first chapter is to represent a message in as efficient or economical a way as possible, subject to the requirements of the devices that are to deal with it. For instance, computer memory stores information in binary form, essentially as strings of 0s and 1s. Everyone knows that English text contains far fewer letters q or j than e or t. So it is common sense to represent e and t in binary by shorter strings than are used for q and j. It is that common-sense idea that we shall elaborate in this chapter.

We do not consider at this stage any devices that corrupt messages or data. There is no error creation, so no need for error detection or correction. We are thus doing *noiseless* coding, and decoding. In later chapters we meet 'noisy' channels, that introduce occasional errors into messages, and will consider how to protect our messages against them. This will not make what we do in this chapter unnecessary, for we can employ coding and decoding for error correction *as well* as the noiseless coding and decoding to be met with here.

The first mathematical idea we shall consider about noiseless coding — beyond just setting up notation, though that carries ideas along with it — is that codes should be *decipherable*. We shall, naturally, insist on that! The mathematical expression of the idea, the *Kraft inequality*, limits how little code you can get away with to encode your messages. Under this limitation you still have much choice of code, and need therefore a criterion of what makes a code optimal. Now the problem is not to encode a single message, but to set up the method of encoding an indefinitely long stream, stretching into the future, of messages with similar characteristics. The likely characteristics of those prospective messages have to be specified probabilistically. That is, there is a message 'source' whose future output *from the point of view of having to code it,* is random, following a particular probability distribution or distributions which can be ascertained from the physical set-up or estimated statistically.

Once the idea of a random source data-stream is accepted, the optimality criterion for a code virtually has to be that on average, in the probabilistic expected-value sense, there should be as little coded message as possible for the amount of source message being encoded. The main theorem of this chapter, the *Noiseless-coding Theorem* of C. E. Shannon, establishes within quite close bounds what sort of expected length the coded messages from an optimal source will have. The bounds involve a certain 'source entropy' which will be the first appearance in this text of the vitally important concept of *mathematical entropy.* We shall consider this more formally and at length in later chapters, but it seems best to introduce it first in a context where it arises naturally in the solution to a particular coding problem.

Other material in the chapter, following the Noiseless-coding Theorem, is a brief consideration of another major idea for later in the book, that of *block* or *segmented* coding (taking large, fixed-length chunks of source output), and then a full treatment of two specific methods for noiseless coding, one quick and easy, suggested by the proof of the Noiseless-coding Theorem, and one optimal.

Most students find the material of this chapter very straightforward, but the sort of exercises that typically go with it are perhaps a little too straightforward. So in the exercises at the end of the chapter we have made an effort to go beyond those.

1.1 Sources, messages, codes

A *source* **U** emits *letters* taken from a finite set F_m of m elements, the *source alphabet.* Its elements are also called *source letters* or *source symbols.* A 'message' will be simply a string of source letters, so we should say precisely what we mean by 'string':

Definitions. *Let C be a set. An n-string of elements of C is $c_1 c_2 \ldots c_n$, where each $c_k \in C$. (That is, an n-string is simply a row n-tuple from C, which we write without commas or spaces where this causes no ambiguity.) The set of all such n-strings is denoted C^n.*

A string of elements of (or just 'from') C is an n-string for some n. The set of all strings from C is denoted C^*.

The length of a string $c = c_1 c_2 \ldots c_n$ *is* $|c| = n$.

So every lower-case word on this page is a string from the 26-letter roman alphabet, while !!***@***!! is an 11-string of punctuation marks, as in your favourite comic strip.

We can now define a *(source) message* to be a string from the source alphabet F_m.

It is to be encoded using an alphabet G_a, a set of a elements, the *code alphabet*. Here usually $a < m$, so each source letter i has to be represented by a *codeword*, a string of elements of G_a. The codewords are thus those strings used, one for each of the m source letters.

You will get the idea best from looking at the following examples of codes that have been used in practice.

Example 1.1.1. The *Greek fire code* is one of the earliest known 'economical representations', described by the writer Polybius* in 208 BC. It was to be used by military commanders to transmit messages from mountain top to mountain top. It codes the 24-letter Greek alphabet into a 5-symbol alphabet, in the obvious way, as is clear from Table 1.1.1.

Table 1.1.1. Greek fire code

Source letter	Codeword	Source letter	Codeword	Source letter	Codeword
α	11	ι	24	ρ	42
β	12	κ	25	σ	43
γ	13	λ	31	τ	44
δ	14	μ	32	υ	45
ε	15	ν	33	ϕ	51
ζ	21	ξ	34	χ	52
η	22	o	35	ψ	53
θ	23	π	41	ω	54

The way the code was used was by displaying flaming torches. One or two torches would be held up at one position on the mountain top, and up to 5 at a position nearby. To transmit letter θ, for instance, 2 torches would be shown at the left-hand position and 3 at the right. If ε was to be transmitted next then 1 torch would be shown at the left-hand position and 5 at the right. Polybius argues that, with practice, this signalling becomes rapid and easy.

Example 1.1.2. Samuel Morse (1791–1872) invented the *Morse alphabet* or code for use with the electric telegraph, with which he is also credited. The 'alphabet'

---

* *Histories*, X, 43–47. Loeb Classical Library ed., vol. IV, pp. 206–219. New York, 1966.

became extensively used for radio communication and can still be heard on short-wave bands today.

Table 1.1.2. The Morse alphabet

Source letter	Codeword	Source letter	Codeword	Source letter	Codeword
a	· —	j	· — — —	s	· · ·
b	— · · ·	k	— · —	t	—
c	— · — ·	l	· — · ·	u	· · —
d	— · ·	m	— —	v	· · · —
e	·	n	— ·	w	· — —
f	· · — ·	o	— — —	x	— · · —
g	— — ·	p	· — — ·	y	— · — —
h	· · · ·	q	— — · —	z	— — · ·
i	· ·	r	· — ·		

In coding by Morse code the raw text is considered to be entirely in (say) lower case. However the inter-word spaces are retained, so the source alphabet has in effect $m = 26 + 1 = 27$ letters. (There are also codes for the numerals 0–9, accented letters and punctuation, but we ignore those.) Likewise the letters of Morse code need spaces to separate them, or confusion obviously results. So the code is in fact *ternary*, having $a = 2 + 1 = 3$ symbols dot, dash and space. Multiple spaces are transmitted to denote the spaces between the words of the source message.

We formally define the notion 'code' as follows.

Definition. *A code is a map f from F_m to G_a^*, the set of all strings of elements of G_a.*

So a message $u_1 u_2 \ldots u_n$ is represented by the *concatenation* of codewords $f(u_1)f(u_2) \ldots f(u_n)$.

A code alphabet of a letters gives an 'a-ary code'. 2-ary, or *binary*, codes are those most often considered.

Codes as we present them in this book are not *secret* codes, but simply transformed messages. The transformations try to ensure only the efficient representation or transmission of the messages, as with the two examples above. They may be hard to 'un-transform', or decode, but that is just an accidental side-effect. The word 'cipher' is sometimes reserved for describing codes specifically designed for secrecy. The science of ciphering, or *cryptography*, needs a book to itself, and we shall not touch on it.

1.2   Decipherability: the prefix condition

Definition. *A code is* decipherable *or* uniquely decodable *if every string from G_a^* is the image of at most one message.*

Example 1.2.1. Consider the codes I, II and III specified in Table 1.2.1.

For each code you have $i \in F_4$ and $f(i) \in G_2^*$, that is, the codes are from the 4-letter alphabet $F_4 = \{1, 2, 3, 4\}$ to strings in the binary alphabet $G_2 = \{0, 1\}$.

Code I is clearly not decipherable, as for instance 0001 is very ambiguous as a coded message. It can be read as $(0)(0)(01)$, corresponding to source message 114, or as $(00)(01)$, corresponding to source message 34, and in three other ways as well!

Table 1.2.1

i	I $f(i)$	II $f(i)$	III $f(i)$
1	0	0	0
2	1	10	01
3	00	110	011
4	01	111	111

Code II is decipherable. There is for instance only one way to decode 10111110. (The answer is the decimal number that represents 3^5.)

Code III is also decipherable. Something like 011111111000001111 is easiest to decode if you start from the right.

So of these three codes, I is useless, II looks worthwhile and III is good only for those who like reading from right to left.

A final point that these codes illustrate is that there can be *illegal* coded messages, that correspond to no source message. In code II the string 011 is one such. This explains why the phrase 'at most one message' appears in the definition of decipherability.

The following notions will clarify what lies behind the differences between the codes in the above example.

Definitions. *A string x is a* prefix *in a string y if there exists a string z such that $xz = y$.*

(For instance in English text 'un' is a prefix in 'unable' but is not a prefix in 'able'.)

A code satisfies the prefix condition *('is a* prefix-free code*') if no codeword is a prefix in any other.*

(For instance code II above is prefix-free but codes I and III are not.)

Fact *(Exercise 1). A prefix-free code is decipherable.*

You can decode the prefix-free code II *as you read the message*, from left to right:

0110111100

↓   ↓   ↓ ↓↓

1   3   4 21

This observation leads straight to the next idea:

Definition. *A code is instantaneous if it is decipherable without lookahead (i.e. a word can be recognized as soon as complete).*

Fact *(Exercise 1). A code satisfies the prefix condition if and only if it is instantaneous.*

Example 1.2.2. Code III above is decipherable but not prefix-free. For only when you read 0 or the end of the message can you decode back to where you had previously decoded. Reading the following message from left to right, you can decode it (as 3444) only when it ends:

$$0\,1\,1\,1\,1\,1\,1\,1\,1\,1\,1\,1$$
$$\underbrace{}_{3}\ \underbrace{}_{4}\ \underbrace{}_{4}\ \underbrace{}_{4}$$

So code III needs an *indefinitely large memory*: the whole message may have to be stored.

Two morals suggest themselves from the above discussion:

- *The prefix condition is sufficient but not necessary for decipherability.*
- *Among decipherable codes the prefix-free codes have extra advantages.*

You will see in the next section that for any decipherable code there exists a prefix-free code with the same word-lengths. So among decipherable codes you need not look beyond the prefix-free codes. This is just as well, because to check a non-prefix-free code for decipherability is quite tricky (see Ash (1965), pp. 29–33 for the algorithm). In effect, though, you never need to do so.

Decision-tree representation

There is a natural way to depict a code by a tree diagram. The tree is made up of nodes and arcs. There is a root node, from which arcs (branches) grow, and at the tips of these branches are other nodes from which further branches grow, and so on. Each branch is labelled by a code letter. For an a-ary code there are at most a branches growing from each node. So for a binary code there are at most 2, and a binary tree results, as in Fig. 1.2.2. We draw the branches growing from each node *upwards* from it.

Each source symbol sits at a node. Its codeword is got by reading off the labels on the arcs that lead from the root to it, in order. You can see that the diagrams for the three codes of Example 1.2.1 are as in Fig. 1.2.2.

We will always assume that each *leaf* (a node with no branches growing from it) has a source symbol sitting at it, so that you draw no more tree than you need to for a given code.

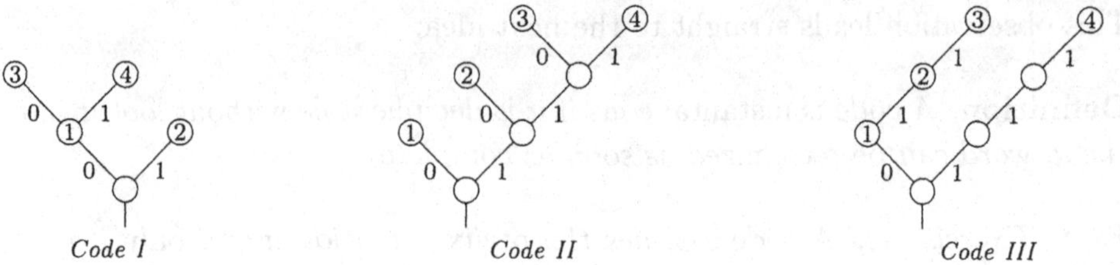

Code I   Code II   Code III

Figure 1.2.2

It is clear then that a code is prefix-free if and only if every source symbol sits at a leaf. In a prefix-free code no source symbol can sit at an *internal* (non-leaf) node.

The tree diagram for a binary prefix-free code is sometimes known as a 'decision tree' because it can be interpreted as directing a question-and-answer session to elicit an unknown source letter. The binary-code symbols 1, 0 are interpreted as 'yes', 'no' answers respectively. You will get the idea from Fig. 1.2.3, which shows the tree of a prefix-free code together with a dialogue directed by it. The speakers are playing a game known as 'Bar-kochba'.*

"Think of a whole number in the range 1, ..., 5."
"OK."
"Is it 2 or 3?"
"No."
"Is it bigger than 1?"
"Yes."
"Is it bigger than 4?"
"No."
"It's 4."
"Right!"

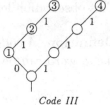

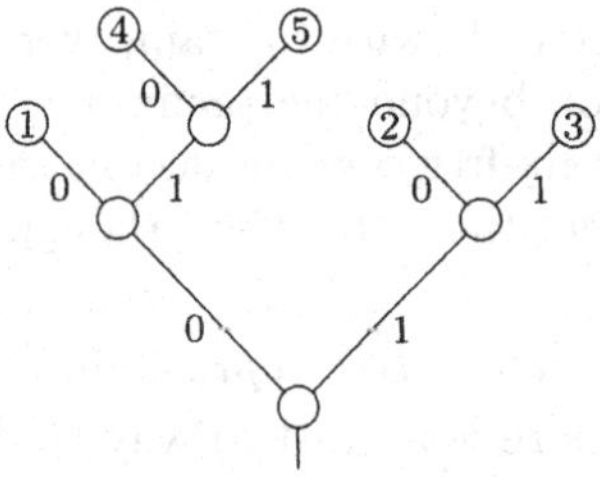

Figure 1.2.3

Exercises

1.   Show that prefix-free codes are decipherable. Show that a code is
       instantaneous if and only if it is prefix-free.
           The *reversal* of a code is obtained by taking all the codewords
       in reverse order. Is the reversal of a decipherable code always a
       decipherable code? What about prefix-free codes?

---

*   which dates back either to Budapest, c. 1900, or to 135 BC. See Rényi (1984), p. 13.

2. A *complete* a-ary tree is one in which every internal node has exactly a branches growing from it. (But note that in graph theory this is called a *regular* a-ary tree, as a 'complete' graph is differently defined.)

    (a) Show that in a complete binary tree the number l of leaves and the number i of internal nodes are related by $l = i + 1$.

    (b) You have decided to use a certain complete binary tree to code your source alphabet, and have allocated the source letters to the m leaves. How many choices of prefix-free code do you now have, i.e. how many proper labellings of the arcs by 0 and 1?

3. Show that an alphabet of m letters can be encoded by an a-ary code with a complete tree as in Exercise 2 if and only if $m \equiv 1$ (mod $a - 1$), that is, $m = 1 + k(a - 1)$ for some integer k.

4. Consider a-ary codewords ordered as numbers to base a in $[0, 1)$, e.g. (for $a = 3$) $00 < 020 < 10$ corresponding to $\cdot 00 < \cdot 020 < \cdot 10$. More formally this is *lexicographic* or *dictionary* ordering: start with G_a ordered, and define $w, w' \in G_a^*$ to have the property $w < w'$ iff $w = vsu$ and $w' = vs'u'$ where $v, u, u' \in G_a^*$, $s, s' \in G_a$ and $s < s'$.

    Show that this is a total order on G_a^*, i.e. for any two different strings w and w' in G_a^* you have either $w < w'$ or $w > w'$.

5. For a code $f : I \to J^*$ and a code $f' : I' \to J'^*$ the *product code* is $g : I \times I' \to (J \cup J')^*$ given by $g(i, i') := f(i)f'(i')$. In other words you code ordered pairs (i, i'), one element from each source alphabet, by simply stringing together their codes under f and f' respectively. Show that the product of two prefix-free codes is prefix-free, but that the product of a decipherable code and a prefix-free code need not even be decipherable.

6. Show that a prefix-free code will uniquely decode the coding
    $$f(u_1)f(u_2)\ldots f(u_n)\ldots$$
    of an infinite source message. Demonstrate that this is not so for general decipherable codes, by finding such a code f and two distinct infinite source messages whose codings coincide.

1.3 The Kraft inequality

Theorem 1.3.1. *A decipherable code in G_a^* with words of lengths s_1, s_2,*

..., s_m exists **iff**

$$\sum_{i=1}^{m} a^{-s_i} \leq 1 \qquad \text{(the Kraft inequality).} \qquad (1)$$

If so, there is a prefix-free code with words of these lengths.

For instance, code I of Example 1.2.1 fails the inequality, as its word-lengths are too short.

Note 1. A code satisfying (1) is not necessarily decipherable.
Note 2. So prefix-free codes suffice.

Proof. Suppose (1) satisfied. Construct a prefix-free code. First rewrite (1) as

$$\sum_{l=1}^{s} n_l a^{-l} \leq 1$$

where n_l is the number of legal words of length l, and $s = \max s_i$ is the maximum word-length. Rewrite again in the form

$$n_s \leq a^s - n_1 a^{s-1} - \cdots - n_{s-1} a.$$

Since $n_s \geq 0$, deduce

$$n_{s-1} \leq a^{s-1} - n_1 a^{s-2} - \cdots - n_{s-2} a.$$

Since $n_{s-1} \geq 0$, deduce

$$n_{s-2} \leq a^{s-2} - n_1 a^{s-3} - \cdots - n_{s-3} a$$

$$\cdots$$

$$\cdots$$

$$n_3 \leq a^3 - n_1 a^2 - n_2 a$$

$$n_2 \leq a^2 - n_1 a$$

$$n_1 \leq a.$$

Code construction: choose n_1 words of length 1, using distinct symbols from G_a.

This leaves $a - n_1$ symbols unused: you can form $(a - n_1)a$ words of length 2 by appending a letter to each.

Choose n_2 words of length 2 from these: that leaves $a^2 - n_1 a - n_2$ prefixes of length 2.

These can be used to form $(a^2 - n_1 a - n_2)a$ words of length 3, from which you choose n_3 arbitrarily, etc.

Continue, every time keeping the new words not having previous words as prefixes. This gets a prefix-free code.

Conversely, suppose we have a decipherable code in G_a^* with word-lengths $s_1, \ldots, s_m$. Set $s := \max s_i$. Choose $r \in \mathbb{N}$, then

$$\left(a^{-s_1} + \cdots + a^{-s_m}\right)^r = \sum_{l=1}^{rs} b_l a^{-l} \quad \text{for some } b_l \in \{0, 1, 2, \ldots\}. \tag{2}$$

Here b_l is the *number of ways r words can be put together in order to make a string of length l.*

For instance, for code II of Example 1.2.1,

$$(2^{-1} + 2^{-2} + 2^{-3} + 2^{-3})^2 = 2^{-2} + 2(2^{-3}) + 5(2^{-4}) + 4(2^{-5}) + 4(2^{-6}),$$

and the term $2(2^{-3})$ represents words 010 and 100.

Because of decipherability, these strings must be distinct. There are in all a^l strings of length l. So $b_l \le a^l$ for all $l = 1, \ldots, rs$. Substitute in (2):

$$\left(\sum_{i=1}^{m} a^{-s_i}\right)^r \le rs,$$

$$\therefore \quad \sum_{i=1}^{m} a^{-s_i} \le r^{\frac{1}{r}} s^{\frac{1}{r}} \to 1 \qquad (r \to \infty). \qquad \square$$

Inequality (1) was proved to be necessary and sufficient for existence of a prefix-free code by L. G. Kraft in his 1949 Master's thesis. B. McMillan extended the necessity half to general decipherable codes in 1956, and the short proof that we give was published in 1961 by J. Karush.

Exercises

1. Show that for any positive integer m there exists a binary prefix-free code whose word-lengths include all the integers 1, 2, ..., m. How many words of each length can there be in the code?

2. In a *comma* code one code-symbol is used to end every codeword and is used in no other way. Suppose, for a-ary encoding of an m-letter alphabet, that the m words consist of the m shortest distinct strings of non-comma code letters followed by a comma. Such a coding is plainly prefix-free. Verify that it satisfies the Kraft inequality.

1.4 Noiseless-coding Theorem: source entropy

We try to find a best decipherable code.

Suppose that, for $i = 1, \ldots, m$, the source emits $u_i \in F_m$ with probability p_i. (We do not assume successive letters independent.) So the emission of a single letter is described by a random variable U with $P(U = u_i) = p_i$. One would expect to estimate the p_i by long-run frequencies.

Encode u_i by a word $f(u_i) \in G_a^*$, of length $s_i := |f(u_i)|$ say. So the random codeword U is encoded by $f(U)$, and we write $S := |f(U)|$ for the length of this random codeword. So S is itself a random variable.

Criterion. *Use a decipherable code that minimizes expected codeword-length $E(S)$.*

We can calculate expected codeword-length as follows:

$$E(S) = E|f(U)|$$
$$= \sum_{i=1}^{m} p_i |f(u_i)|$$
$$= \sum_{i=1}^{m} p_i s_i.$$

Thus, concentrating just on the lengths of the codewords, rather than on the actual code f, we are led to the following.

Problem. *Find $s_1, \ldots, s_m$ to*

$$\text{minimize} \quad \sum_{i=1}^{m} p_i s_i$$

$$\text{subject to} \quad s_i \in \{0, 1, 2, \ldots\} \tag{1}$$

$$\text{and} \quad \sum_{i=1}^{m} a^{-s_i} \leq 1. \tag{2}$$

Inequality (2) is of course the decipherability condition (the Kraft inequality). An exercise at this point (Exercise 1) is to use a Lagrange-multiplier method to show that, **ignoring** (1) but assuming all $p_i > 0$, the solution to the above constrained-minimization problem is $s_i := -\log_a p_i$. In other words, take the s_i so that $p_i = a^{-s_i}$.

That would give

$$\inf E(S) = -\sum_i p_i \log_a p_i = \frac{-\sum_i p_i \log_2 p_i}{\log_2 a}.$$

This is the first appearance of the (instantaneous) *source entropy* (or 'information measure')

$$h = h(U) := -\sum_{i=1}^{m} p_i \log_2 p_i.$$

The qualification 'instantaneous' is inserted because h is so far based only on single letters from the source. Later we shall remove this restriction.

You should note that h depends only on the probability distribution $p = (p_i)$ of the random variable U, not on its values u_i, so we will also write $h(p)$ when the distribution p is of interest.

With constraint (1) this infimum becomes in general unattainable, but remains a lower bound. We will now prepare to give a proof of this bound without using Lagrange multipliers, and to see how close we can get to it.

$$\boxed{\log \equiv \log_2, \ \ln \equiv \log_e \quad \text{henceforth.}}$$

Lemma 1.4.1. *Let $p = (p_i)$, $q = (q_i)$ be probability distributions on some finite set. Then*

$$-\sum_i p_i \log p_i \le -\sum_i p_i \log q_i \qquad \textit{(Gibbs inequality)}$$

with equality iff $p = q$.

(Interpret the summand as 0 if $p_i = 0$.) The same statement holds for any logarithm.

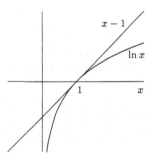

Figure 1.4.1

Proof. On multiplying through by $\ln 2$ you get the same inequality with log replaced by ln. So we may equivalently prove $\sum_i p_i \ln(q_i/p_i) \le 0$. Now $\ln x \le x - 1$ for all $x > 0$, with equality if and only if $x = 1$, as Fig. 1.4.1 illustrates. Let $\mathcal{I} := \{i : p_i > 0\}$, then

$$\sum_i p_i \ln \frac{q_i}{p_i} = \sum_{\mathcal{I}} p_i \ln \frac{q_i}{p_i}$$

$$\le \sum_{\mathcal{I}} p_i \left(\frac{q_i}{p_i} - 1 \right)$$

$$= \sum_{\mathcal{I}} q_i - \sum_{\mathcal{I}} p_i$$

$$= \sum_{\mathcal{I}} q_i - 1 \le 0.$$

For equality you need $\sum_{\mathcal{I}} q_i = 1$, so $q_i = 0$ when $p_i = 0$; you also need $q_i/p_i = 1$ for $i \in \mathcal{I}$. So in fact you need $q_i = p_i$ for all i. $\square$

It is very convenient to have a label to attach to the inequality of the above Lemma. In naming it as we do we follow Hamming (1980), p. 117, though we suspect the connection with the physicist Gibbs is remote at best.

Theorem 1.4.2: Noiseless-coding Theorem (Shannon). *For a random source emitting $u_i \in F_m$ with probability $p_i > 0$, where $i = 1, \ldots, m$ and*

$m \geq 2$, the minimal expected word-length for a decipherable (letterwise) encoding in an alphabet of $a \geq 2$ symbols satisfies

$$\frac{h}{\log a} \leq E(S) < \frac{h}{\log a} + 1 \qquad (3)$$

(recalling $\log \equiv \log_2$). The lower bound is attained iff $p_i = a^{-s_i}$ for **integers** s_i.

Proof. For the left-hand inequality,

$$
\begin{aligned}
h - E(S) \log a &= -\sum p_i (\log p_i + s_i \log a) \\
&= \sum p_i \log(a^{-s_i}/p_i) \\
&= \sum p_i \log \frac{p_i^*}{p_i} + \log \sum a^{-s_i} \qquad \text{where } p_i^* = \frac{a^{-s_i}}{\sum_j a^{-s_j}} \\
&\leq 0 + 0 = 0
\end{aligned}
$$

by the Gibbs and Kraft inequalities. Equality holds iff $p_i = p_i^*$ and $\sum a^{-s_i} = 1$, that is, $p_i = a^{-s_i}$.

For the right-hand inequality of (3), let s_i be the integer such that

$$a^{-s_i} \leq p_i < a^{-s_i+1}.$$

The weak inequality here implies the Kraft inequality is satisfied so a decipherable code with these s_i exists. From the strict inequality,

$$s_i < -\frac{\log p_i}{\log a} + 1$$

so

$$E(S) < \frac{-\sum p_i \log p_i}{\log a} + \sum p_i,$$

whence the right-hand inequality of (3). $\qquad \square$

For a source alphabet of size 2^k, with equally likely letters ('equidistribution'), you calculate $h = -\sum_{i=1}^{2^k} 2^{-k} \log 2^{-k} = k$. The natural way to encode such a source is with all codewords the same length, and you thus need all binary words of k *bits* (binary digits) to do it. So *the units of h are bits.*

Exercises

1.   Use a Lagrange multiplier to solve the problem posed early in the section: with the $p_i > 0$ given, such that $\sum_1^m p_i = 1$, find real numbers $s_1, \ldots, s_m$ to minimize $\sum_{i=1}^m p_i s_i$ subject to $\sum_{i=1}^m a^{-s_i} \leq 1$.

2.   Prove that $\ln x \leq x - 1$ for all $x > 0$.

3.   Consider extending the Noiseless-coding Theorem to cases where some of the p_i are zero. The set-up is thus that there are one or more letters in the source alphabet which are emitted with probability 0, but which you are nonetheless to code.

(a) Show that the result of the theorem, inequality (3), is violated if the source is *degenerate*: $p_i = 1$ for some i.

(b) Show that (3) does extend to general *non-degenerate* sources.

4.   Prove that a probability distribution $p = (p_i)_{i=1,...,m}$ on an alphabet of m letters has $0 \le h(p) \le \log m$, and characterise the cases of equality.

5.   You are given m coins, one of which may be a forgery. Forged coins are either too light or too heavy. Let the presence of a forgery and whether, if present, it is light or heavy, be described by a probability distribution. Use the result of the previous exercise to find the maximum entropy this distribution can have.

    You are also given a balance on which you can place any of the coins that you wish. The coins placed in either pan may be together heavier or lighter than those in the other pan, or the pans may balance. Find the maximum entropy for a probability distribution over these three eventualities.

    You are allowed at most 3 uses of the balance. Deduce that if $m > 13$ you cannot be sure of detecting the forgery and its nature.

6.   For a source with letter-distribution $p = (p_i)_{i=1,...,m}$, the *redundancy* of a binary encoding is $E(S) - h(p)$. Prove that for each ε with $0 \le \varepsilon < 1$ there exists a p such that the *optimal* encoding has redundancy ε.

    So the upper bound in the Noiseless-coding Theorem is 'sharp': it cannot in general be lowered.

7.   Suppose that a gastric infection is known to originate in exactly one of m restaurants, the probability it originates in the jth being p_j. A health inspector has samples from all the m restaurants and by testing the pooled samples from a set A of them can determine with certainty whether the infection originates in A or its complement. Let $N(p)$ denote the minimal expected number of such tests needed to locate the infection. Show that $h(p) \le N(p) \le h(p) + 1$, and determine when the lower bound is attained.

1.5 Segmented (block) codes

Take the source message in *segments* of length n letters: code these. The source considered in this way is the nth-*order extension* $\mathbf{U}^{(n)}$ of the original source $\mathbf{U}$. The random segment $U^{(n)}$ has a probability distribution over

$$F_m^n = \left\{ \begin{pmatrix} u_1 \\ \vdots \\ u_n \end{pmatrix} : u_i \in F_m \right\} = F_m \times F_m \times \cdots \times F_m$$

with entropy $h_n = h(U^{(n)})$. If we code the segments by some code $f_n : F_m^n \to G_a^*$ then the encoded word is $f_n(U^{(n)})$, with length $S^{(n)} = |f_n(U^{(n)})|$. The Noiseless-coding Theorem tells us for an optimal code that $h_n/\log a \le E(S^{(n)}) \le h_n/\log a + 1$, whence

$$\frac{h_n}{n \log a} \le E(S^{(n)}/n) < \frac{h_n}{n \log a} + \frac{1}{n}. \tag{1}$$

This is an important inequality, bounding $E(S^{(n)}/n)$ which is the minimal expected word-length of an encoded message on a *per-source-letter basis*. Its units are *bits/symbol*.

If the source letters are *independent* (a *Bernoulli source*) then $h_n = nh$ (Exercise 1). So in this Bernoulli case $E(S^{(n)}/n) \to h/\log a$ as $n \to \infty$, and you can get as close to the lower bound as you wish by taking n large. There is thus an advantage in segmented codes, to be balanced against the extra complication and storage requirements needed to deal with the source data-stream segmentally.

Example 1.5.1. Consider a binary Bernoulli source with alphabet $\{A, B\}$. Successive letters are emitted independently of one another according to the probability distribution $p_A = p$, $p_B = 1 - p$. Thus $m = 2$ and

$$h = -p \log p - (1 - p) \log(1 - p).$$

We shall encode the source output in binary, so also $a = 2$. Suppose $p = \frac{4}{5}$. According to the calculations above,

$$\frac{h_n}{n \log a} = \frac{h}{\log a} = h = 0 \cdot 72 \text{ bits/symbol.}$$

We investigate how close we can bring $E(S^{(n)}/n)$ to this value by taking the source output in segments of length 1, 2 and 3. It is not important at this stage how the codes were obtained; they are in fact optimal (Huffman) codes, as you will see in §1.7. But the point here is that codes *exist* that bring $E(S^{(n)}/n)$ close to the lower bound h simply by the device of coding in threes.

Code in 1s:

Segment	Code	Probability	
A	0	0·8	$E(S^{(n)}/n) = 1$
B	1	0·2	

Code in 2s:	Segment	Code	Probability
	AA	0	0·64
	AB	10	0·16
	BA	110	0·16
	BB	111	0·04

$$E(S^{(n)}/n) = \frac{1·56}{2} = 0·78$$

Code in 3s:	Segment	Code	Probability
	AAA	0	0·512
	AAB	100	0·128
	ABA	101	0·128
	BAA	110	0·128
	ABB	11100	0·032
	BAB	11101	0·032
	BBA	11110	0·032
	BBB	11111	0·008

$$E(S^{(n)}/n) = \frac{2·184}{3} = 0·728$$

The above demonstrates that by coding segmentally you can reduce as much as you want the discreteness effect that creates the unit gap between the Noiseless-coding Theorem's bounds. As (1) shows, for general, not just Bernoulli, sources you can bring the mean per-source-letter codeword-length close to $h_n/(n \log a)$. We will show later (§2.11) that there is a much greater advantage in segmented codes, in that in quite general *stationary* cases h_n/n is non-increasing. For sources with a lot of 'redundancy', where not much content is being conveyed relative to the quantity of source letters, h_n/n and hence $E(S^{(n)}/n)$ can decrease substantially. Even if $m = a$ one can well have $h_n/(n \log a) < 1$, so that $E(S^{(n)}/n) < 1$ for large n, that is, *data compression*.

Exercises

1. For a Bernoulli source **U** you have for the nth-order extension that $h_n = - \sum_{i_1,...,i_n} p_{i_1} \cdots p_{i_n} \log(p_{i_1} \cdots p_{i_n})$. Show that $h_n = nh$.

2. A Bernoulli source emits 0, 1 with equal probability. Output is to be encoded by segments into the code alphabet A, B, C, D in such a way that $E(S^{(n)})/n < 1$. Find a suitable n, and a code.

3. The representation known as binary-coded decimal encodes 0 as 0000, 1 as 0001 and so on up to 9, coded as 1001, with other 4-digit binary codes being discarded. Why is this code decipherable?
    Consider a Bernoulli source that emits equally likely digits in the range 0–9. It thus has entropy log 10. Consider encoding analogously in blocks, i.e. coding all k-blocks of decimal digits by binary m-blocks. Prove that for suitable k and m you can get arbitrarily

near the lower bound $\log 10$ on codeword-length per decimal digit. (Cambridge 1982)

1.6   Shannon-Fano encoding

This is as in the proof of the Noiseless-coding Theorem: word-lengths s_i, *integers* such that $a^{-s_i} \le p_i < a^{-s_i+1}$, so that

$$- \log_a p_i \le s_i < - \log_a p_i + 1,$$

and thus $s_i := \lceil - \log_a p_i \rceil$ (where $\lceil x \rceil$ is the least integer $k \ge x$). Then construct a prefix-free code from the shortest s_i upwards, taking the binary numbers in order, ensuring that previous codes are not prefixes. The Kraft inequality guarantees enough room.

For a programmable algorithm, take for instance the binary expansion, to s_i places, of $\sum_{j=1}^{i-1} p_j$. Remove the initial dot and take $f(i)$ to be the s_i-string remaining.

Example 1.6.1. Here is a case with $m = 5$ and $a = 2$. The 5-letter alphabet with probabilities shown in Table 1.6.1 is to be coded in binary. The formulae above give the word-lengths s_i to be used. There are a number of easy ways to write down a prefix-free code with these word-lengths, but we have chosen here to illustrate the algorithm mentioned above. Thus for source-symbol 2 you have $\sum_{j=1}^{i-1} p_j = p_1 = 0\cdot4$. Expanded in binary to 3 places this is $\cdot011$, so the symbol is coded by 011.

Table 1.6.1

i	p_i	$-\log p_i$	s_i	$f(i)$
1	0·4	1·32	2	00
2	0·2	2·32	3	011
3	0·2	2·34	3	100
4	0·1	3·32	4	1100
5	0·1	3·32	4	1110

A code with these word-lengths has $E(S) = 2\cdot8$, whereas $h = 2\cdot12$. It is clear that the Shannon-Fano recipe can be rather wasteful, in that for instance here the last digit can be dropped from the two longest codewords.

Exercise

1.   Verify that Shannon's method of coding, taking $f(i)$ to be the binary expansion to s_i places of $\sum_{j=1}^{i-1} p_j$, does give a prefix-free code. Use it to code a source with probability distribution $\cdot22$, $\cdot2$, $\cdot18$, $\cdot15$, $\cdot1$, $\cdot08$, $\cdot07$.

1.7   Huffman encoding

This is an algorithm to produce an *optimal* coding from F_m to G_a^* (i.e. finite sequences from G_a), where $a \le m$. We give the algorithm in the binary case $a = 2$ only.

The method is to construct the tree of a *prefix-free* code from the top, as follows.

(i) Order the letters of F_m so that $p_1 \ge p_2 \ge \cdots \ge p_m$.
(ii) Assign '0' to $m - 1$, '1' to m.
(iii) Now start again from the *reduced* alphabet $F_{m-1} = \{1, 2, \ldots, m - 2, (m-1, m)\}$ with respective probabilities $\{p_1, p_2, \ldots, p_{m-2}, p_{m-1} + p_m\}$.

Example 1.7.1. We use the Huffman algorithm to encode the 5-letter source with probabilities given in Fig. 1.7.1 into binary.

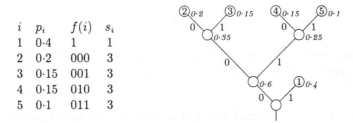

i	p_i	$f(i)$	s_i
1	0·4	1	1
2	0·2	000	3
3	0·15	001	3
4	0·15	010	3
5	0·1	011	3

Figure 1.7.1

The easiest way to do it is via the tree. Think of the probabilities as 'weights' of nodes. The two lowest weights are 0·1 and 0·15, so the nodes for 4 and 5 are drawn first and joined to form a new node of weight 0·25. Of the four weights now to be considered the lowest are 0·15 and 0·2, so the corresponding nodes 2 and 3 are joined. The three weights present now are 0·25, 0·35 and 0·4, so the nodes with the two lowest weights are joined. There are then two nodes, with weights 0·4 and 0·6, and joining these gets the root node, completing the tree. Finally, label the two branches growing from each intermediate node with a 0 and a 1.

We gave for definiteness a rule, (ii), for allocating labels 0 and 1 to the two branches at each cycle of the Huffman procedure. But you do not have to do it that way round: so long as you label one branch '0' and the other '1' an optimal tree and code results. The *tree* is in any case unaltered by the labelling, and in the above example there is clearly a unique tree(-shape) resulting from the algorithm. But when at any stage of the algorithm some of the weights of the nodes *coincide* there can be various distinct trees possible, all equally optimal by the criterion we use. The next example gives an instance (and see Problem 1.9.13). In Problem 1.9.14 you are asked to consider other properties of the optimal trees, such as variance of codeword-length, that could be used to provide refinements of the optimality criterion.

Example 1.7.2. The Huffman algorithm applied to the source of Example 1.6.1 gives the tree and code shown in Fig. 1.7.2.

i	p_i	$f(i)$	s_i
1	0·4	00	2
2	0·2	10	2
3	0·2	11	2
4	0·1	010	3
5	0·1	011	3

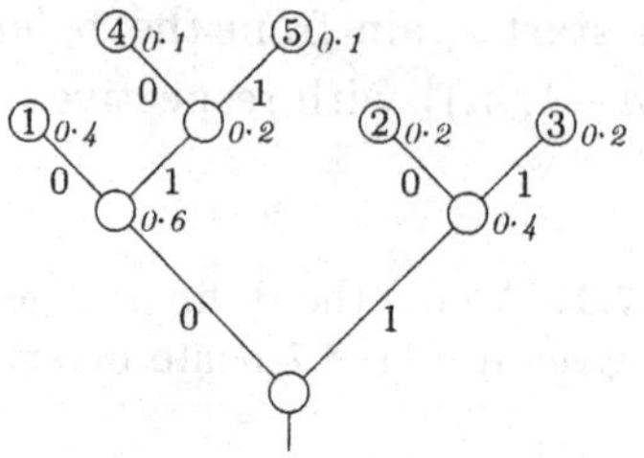

Figure 1.7.2

The average codeword-length is $E(S) = 2\cdot2$, appreciably better than the value 2·8 for the Shannon-Fano code. Recall also that $h = 2\cdot12$ for this source, so the lower bound of the Noiseless-coding Theorem cannot be attained.

To prove Huffman optimal, lemmas are needed giving *necessary* properties of an optimal code:

Lemma 1.7.3. *In an optimal prefix-free code the word-lengths are reverse-ordered with respect to probabilities, viz.*

$$p_i > p_j \quad implies \quad s_i \leq s_j.$$

Proof. If not, form a new code by exchanging the codes of i and j. This shortens expected word-length, and preserves prefix-free. □

Lemma 1.7.4. *In an optimal prefix-free code there must among the words of maximum length be at least 2 agreeing in all but the last digit.*

Proof. Suppose not: so either there is but 1 word of maximum length, *or* there are 2 or more such words, and they differ before the last digit. In both cases you can drop the last digit from all words of maximum length, without losing the prefix condition. □

You do not, in fact, ever have to check a code for the properties in Lemmas 1.7.3–4, as the Huffman construction gives codes that are optimal and so have these properties.

Theorem 1.7.5. *Among decipherable codings Huffman is optimal (has minimal expected word-length).*

Proof. (Binary codes only: $a = 2$.) If $m = 2$ the Huffman code is 0, 1, which is optimal. So assume inductively that Huffman is optimal for F_{m-1}, whatever the probability distribution. To prove the Huffman code f_m for F_m is optimal, suppose not, that is that there exists another code f_m^* for F_m with *shorter* expected word-length; $E(S_m^*) < E(S_m)$. Denote the probabilities by $p_1 \geq \cdots \geq p_m$. In both codes you can shuffle words of the *same* length, so may alter both if need be so that, by the lemmas, the words corresponding to p_{m-1} and p_m have maximum length and differ only in their final digit.

Now reduce *both* codes by the Huffman method, i.e.

- remove the final digit from $f_m(m-1)$ and $f_m(m)$, getting the Huffman code f_{m-1} for F_{m-1};
- remove the final digit from $f_m^*(m-1)$ and $f_m^*(m)$, getting a prefix-free code f_{m-1}^* for F_{m-1}.

In f_m the contribution of $f_m(m-1)$ and $f_m(m)$ to $E(S_m)$ is $s_m(p_{m-1}+p_m)$, and after the reduction they contribute $(s_m - 1)(p_{m-1} + p_m)$, so $E(S_m)$ is reduced by $p_{m-1} + p_m$.

You get the *same* reduction in $E(S_m^*)$.

We assumed $E(S_m^*) < E(S_m)$, so deduce $E(S_{m-1}^*) < E(S_{m-1})$, that is, the new code f_{m-1}^* for F_{m-1} has shorter expected word-length than the Huffman code f_{m-1}, a contradiction. $\qquad\square$

Huffman encoding in general, with a code alphabet of size a, is not hard once you are used to the algorithm in the binary case, and you are invited to explore the possibilities in the Exercises below and in Problems 1.9.11, 1.9.15 and 1.9.16.

Exercises

1.   Find an optimal binary code for the source in Exercise 1.6.1.

2.   (a) Find binary Huffman codes for sources on 2, 3, 4, 5, 6 and 7-letter alphabets, with all letters equally likely.
      (b) Establish what the word-lengths are, and how many times each length appears, for a binary Huffman encoding of a source with an m-letter alphabet of equally likely letters.

3.   Find a *ternary* Huffman encoding for the source in Exercise 1.6.1,

always grouping the 3 lowest probabilities together. What would happen if you used this procedure on an 8-letter source?

4.   Huffman a-ary encoding creates a complete tree (Exercise 1.2.2), as you always combine the leaves together in sets of size a. So it works without adjustment, according to Exercise 1.2.3, only when $m \equiv 1$ (mod $a - 1$). When that is not so the trick is first to adjoin extra leaves, of weight 0, to bring the number of leaves up to $1 + k(a - 1)$ for some integer k. Carry this out for ternary encoding of a source with distribution ·2, ·2, ·15, ·15, ·1, ·1, ·05, ·05.

5.   Construct a quaternary ('4-ary') Huffman code for the following source-alphabet; the probability of each letter is given after it.

    A 0·11 B 0·08 C 0·05 D 0·09 E 0·16 F 0·07
    G 0·09 H 0·04 I 0·18 J 0·07 K 0·06

(Cambridge 1975)

6.   Consider a-ary Huffman encoding of a source with probability distribution $(p_i)_{i=1,...,m}$. Show that for fixed a the number of operations required, as a function of m, is bounded by Km^2 for some K. ('Operations' are addition, comparison, insertion, deletion, attaching a label, ...)

    Now consider the number of operations as a function both of m and of a. Show that Cm^2/a is a bound, for suitable C.

7.   Write a computer program to do binary Huffman encoding. If you know the Pascal language you might do it by setting up a representation of the code's tree using pointers and suitable data structures.

1.8 Further topics

Run-length constraints

Because of the limitations inherent in physical devices, such as those used in magnetic recording, there is a need to perform noiseless coding under 'run-length' constraints, of which the following are typical:

(a)   no more than r consecutive 0s, or 1s;

(b)   at least a 0s follow every 1, but no more than b consecutive 0s allowed;

(c)   the running difference between the total number of 0s issued and the total number of 1s must lie within a certain range.

Constraint (a) is needed by systems that use $0 \to 1$ or $1 \to 0$ transitions to keep parts of the system in synchronization with each other, as well as communicating them. An older instance is that in Morse telegraphy no two consecutive letter-spaces or word-spaces are permitted. Constraint (b) is needed for example by systems with dead-time following a 1. Constraint (c) arises when the 0s and 1s are actual positive and negative charges. It is therefore called a *charge constraint*.

A coding device subject to such a constraint moves between *states*. For a simple example consider a binary stream with at most 2 consecutive 0s or 1s. The states are then

$$S_1 = 0, \ S_2 = 00, \ S_3 = 11, \ S_4 = 1,$$

where being in S_1 means that a single 0 has just been emitted, preceded by 1, and similarly for the other states. The matrix whose $(i,j)^{\text{th}}$ entry is the number of possible transitions from state i to state j is

$$\begin{pmatrix} 0 & 1 & 0 & 1 \\ 0 & 0 & 0 & 1 \\ 1 & 0 & 0 & 0 \\ 1 & 0 & 1 & 0 \end{pmatrix}.$$

Because of the constraint you cannot code an arbitrary binary sequence as fast as it comes in. There is a theorem that says the code will allow you to output on average at most $\log \lambda$ bits for every bit of input, where λ is the largest positive eigenvalue of the matrix. Here $\lambda = (1 + \sqrt{5})/2$ so $\log \lambda \simeq 0{\cdot}694$, and the code will be $1/0{\cdot}694 = 1{\cdot}44$ as long as the uncoded message.

We have implicitly assumed the code symbols take the same time as each other to transmit. An extension of the above approach handles cases where that is not so.

One can attempt to find a good prefix-free code satisfying the run-length constraint. The method is to identify some of the states, called *terminal* states, from which starting positions all of the codewords are legal continuations. Then each codeword should return to a terminal state. In the above example states S_1 and S_4 are obvious terminal states, and a suitable legal prefix-free code for an equally likely 6-letter alphabet might be

$$0101 \quad 01001 \quad 0110 \quad 1010 \quad 10110 \quad 1001.$$

For further details see Zehavi & Wolf (1988), Blake (1982) and Blahut (1987), Ch. 2.

Universal coding

This is the name given to a class of coding methods that do not assume, as does Huffman coding, knowledge of the probability distribution of the source output. Yet the methods are close to optimal, at least for sufficiently large block-sizes.

The simplest case is the encoding of a Bernoulli source with m-letter alphabet, in binary, say. The source output is taken in segments of length n. We do not know the probability distribution of the source output but can estimate it by the vector of relative frequencies $(n_1/n, n_2/n, \ldots, n_m/n)$. Here n_i is the number of times the source letter i appears in the n-segment, so the non-negative integers n_1, ..., n_m have sum n. There are $\binom{n+m-1}{m-1}$ different possible relative-frequency vectors. We can thus encode a list of them by a binary block code of $\lceil \log \binom{n+m-1}{m-1} \rceil$ bits. (Recall the 'ceiling' function $\lceil \ \rceil$ from §1.6.)

Now the number of different n-segments that have a given frequency vector $(n_1/n, \ldots, n_m/n)$ is given by the multinomial coefficient $n!/\prod_1^m n_i!$. We can encode a list of these n-segments by a code of $\lceil \log(n!/\prod_1^m n_i!) \rceil$ bits. The final codeword for the n-segment is then the concatenation of the code for the frequency vector followed by the code for the specific n-segment with that frequency vector, and has length $\lceil \log \binom{n+m-1}{m-1} \rceil + \lceil \log(n!/\prod_1^m n_i!) \rceil$ bits.

For further reading about universal coding see Davisson (1973), Davisson *et al.* (1981), Rissanen (1984). Widely available software on modern computer systems uses dynamically updated universal coding to compress and uncompress files. The 'pkzip' and 'pkunzip' programs for IBM PCs are one version, the 'compress' utility in Unix another. See Welch (1984) for some details.

Tree codes

These make interesting alternatives to Huffman codes as they avoid the need to code in segments to overcome discreteness effects. They operate not with a set of fixed codes for source letters or blocks, but instead allow the source to pick its way down an infinite decision tree, generating code symbols as choices of branches are made. The *Elias code* is mathematically very simple, as follows. Consider a binary Bernoulli source emitting symbol 0 with probability $p \in (0,1)$ and symbol 1 with probability $1 - p$. It is to be considered as doing so into the indefinite future, producing an infinite sequence U_1, U_2, ... of random variables taking values in $F_2 = \{0,1\}$. Let $F_2^{\mathbb{N}}$ denote the set of infinite sequences of elements of F_2, written as infinite 'strings' $\mathbf{u} = u_1 u_2 u_3 \ldots$. So $\mathbf{U} = U_1 U_2 U_3 \ldots$ is a 'random element' of

F_2^N. We may order F_2^N lexicographically, similarly to Exercise 1.2.4. Define $\mathbf{u} = u_1 u_2 \ldots$ and $\mathbf{u}' = u_1' u_2' \ldots$ to have the property $\mathbf{u} < \mathbf{u}'$ if there exists $j \geq 1$ with $u_j < u_j'$, and $u_i = u_i'$ for all $i < j$ (if any). Say that $\mathbf{u} \leq \mathbf{u}'$ if $\mathbf{u} < \mathbf{u}'$ or $\mathbf{u} = \mathbf{u}'$. Now define a real-valued function x on F_2^N by letting $x(\mathbf{u})$ be the probability that the random $\mathbf{U}$ comes at or before $\mathbf{u}$ in lexicographical order:

$$x(\mathbf{u}) := P(\mathbf{U} \leq \mathbf{u}) \qquad (\mathbf{u} \in F_2^N).$$

The code $f(\mathbf{u})$ is the *ordinary binary expansion*, with the initial point removed, of the real number $x(\mathbf{u})$.

In order for the code to be usable there has to be a way of building up the successive digits of $f(\mathbf{u})$ from knowledge of the initial digits of $\mathbf{u}$. A simple recursive scheme suffices. Since $\mathbf{u}$ lies in lexicographical order between $u_1 u_2 \ldots u_n 000 \ldots$ and $u_1 u_2 \ldots u_n 111 \ldots$, $x(\mathbf{u})$ lies as a real number between $x_n = x(u_1 u_2 \ldots u_n 000 \ldots)$ and $x_n + d_n = x(u_1 u_2 \ldots u_n 111 \ldots)$. Now it is easy to see that x_n and d_n are generated by initial values $x_0 := 0$, $d_0 := 1$ and recursions

$$d_{n+1} := \begin{cases} pd_n & \text{if } u_{n+1} = 0, \\ qd_n & \text{if } u_{n+1} = 1, \end{cases}$$
$$x_{n+1} := \begin{cases} x_n & \text{if } u_{n+1} = 0, \\ x_n + pd_n & \text{if } u_{n+1} = 1. \end{cases}$$

Knowledge of these bounds on $x(\mathbf{u})$ enables its binary expansion's initial digits to be found. Use of rational arithmetic avoids rounding errors, which would of course propagate fatally.

As an example, suppose $p = 0 \cdot 8$ and the source output starts 0001000000, then the above bounds for the case $n = 10$ are $\cdot 4096 \leq x \leq \cdot 4364435456$. These imply $\cdot 40625 \leq x \leq \cdot 4375$ which says that the binary expansion of x is between $\cdot 01101$ and $\cdot 01110$. So the first 3 code digits are 011. The source has entropy approximately $0 \cdot 7219$ and the code will compress the source stream in that proportion in the long run.

For further reading on the Elias code see Jones (1981), and for tree codes in general Rissanen (1976).

Other reading

For entropy developed axiomatically, as in Problem 6 below, see Ash (1965), Rényi (1970) or Pinsker (1964).

For code-based questionnaires, search strategies, etc., see Picard (1980), Aigner (1988) and Knuth (1973), §§6.2–6.3.

1.9 Problems

1.   Re-work Exercise 1.2.2 for complete a-ary trees: find (a) a relation between the numbers of internal nodes and leaves, and (b) how many prefix-free codes correspond to a given tree with the source letters already assigned to the leaves.

2.   Let us call an a-ary prefix-free code *full* if every 'infinite string' $a_1 a_2 \ldots$, where the $a_k \in G_a$, can be decoded. Show that the following are equivalent:
(i) the code is full,
(ii) its tree is complete (see Exercise 1.2.2),
(iii) you cannot adjoin an extra codeword to the code without violating the prefix condition,
(iv) $\sum_{i=1}^{m} a^{-s_i} = 1$,
(v) there are no indecipherable strings. (A string in G_a^* is *indecipherable* if it is not a prefix in any finite concatenation of codewords.)

3.   An alphabet of m letters is to be encoded by codewords whose first letters are taken from an alphabet G_1, of $a_1 < m$ letters. Any codewords of length greater than 1 are to have their second letters taken from an alphabet G_2, of a_2 letters. Any codewords of length greater than 2 are to have their third letters taken from an alphabet G_3, of a_3 letters, and so on. The alphabets $G_1, G_2, \ldots$ are given, and are nested: $G_1 \subseteq G_2 \subseteq \ldots$. The standard set-up for noiseless coding is thus the special case when the alphabets $G_1, G_2, \ldots$ coincide.

   Prove that, as in the standard set-up, the prefix condition implies decipherability.

   Prove that for any positive integers $s_1, \ldots, s_m$ satisfying

$$\sum_{i=1}^{m} \prod_{j=1}^{s_i} \frac{1}{a_j} \leq 1, \tag{1}$$

a prefix-free code exists with word-lengths $s_1, \ldots, s_m$.

   Give an example to show that (1) is however not necessary for existence of a decipherable code with word-lengths $s_1, \ldots, s_m$. (Cambridge Dipl. Stat. 1989)

4.   Suppose that an optimal binary code has been constructed using probability distribution $p = (p_i)_{i=1,\ldots,m}$, estimated from frequency data, but that the true distribution is $p' = (p'_i)_{i=1,\ldots,m}$ where $p_i = p'_i + d_i$ for all i.

(a) Show that the true expected codeword-length $E'(S)$ and the value $E(S)$ calculated on the basis of the estimates p_i are related by

$$E'(S) = E(S) - \sum_1^m d_i s_i.$$

(b) Since p and p' are probability distributions you have

$$\sum_1^m d_i = 0. \tag{2}$$

A reasonable summary of the closeness of p and p' is the average squared deviation

$$v_d = \frac{1}{m} \sum_1^m d_i^2. \tag{3}$$

For a given v_d one can ask how far $E'(S)$ and $E(S)$ can be apart. Set $v_s := m^{-1} \sum_{i=1}^m (s_i - \bar{s})^2$ where $\bar{s} = m^{-1} \sum_1^m s_i$. Prove that, under (2) and (3),

$$|E'(S) - E(S)| \le \sqrt{v_s v_d},$$

and that the bound is attainable.

5. You are playing bridge with a partner and two opponents. The pack, of 52 cards, is dealt. A simple representation for the hand you get would assign a unique 6-bit binary number to represent each card; then a 78-bit message represents your hand, a 156-bit message your pair's hands and a 312-bit message the whole deal. Let us do better. Assume all possible deals are equally likely.

(a) Show that there are $52!/(13!39!)$ different hands you might obtain. Deduce that no representation of your hand can use less than about 40 bits. Give a representation using 52 bits.

(b) Show that there are $52!/(13!)^4$ different deals. Deduce that no representation of the deal can involve less than about 96 bits. Give a representation involving 102 bits.

(c) Show that no representation of your and your partner's hands can involve less than about 73 bits. Give a representation involving 78 bits.

6. Consider a 'measure of information' (uncertainty, entropy) H. For any probability distribution p on a finite set $H(p)$ is to be well-defined as a non-negative number. In an axiomatic treatment, the basic property usually demanded of such a measure is that, for any distributions $p = (p_i)_{i=1,\ldots,m}$ and $q = (q_i)_{i=1,\ldots,n}$,

$$H(p_1 q_1, p_1 q_2, \ldots, p_1 q_n, p_2, p_3, \ldots, p_m) = H(p) + p_1 H(q). \tag{4}$$

That is, that if one of the contingencies (of probability p_1) is divided into sub-contingencies, of conditional probabilities q_1, q_2, ..., q_n, then the total uncertainty breaks up additively as shown. The function H is assumed to be symmetric in its arguments, so that analogous relations hold if contingencies 2, 3, ..., m are sub-divided.

Suppose that $h(m) := H\left(\frac{1}{m}, \ldots, \frac{1}{m}\right)$ is non-decreasing in m. Show that, as a consequence of (4),

$$h(m^k) = kh(m).$$

By considering *large* arguments, deduce that

$$h(m) = c \log m$$

where $c = h(2)$.

Hence show, finally, that

$$H(p_1, \ldots, p_m) = -c \sum_{1}^{m} p_j \log p_j$$

if the p_j are rational. The validity of this formula for all p then follows if H is assumed continuous.

7.   Suppose that a source emits letters from the alphabet $\{1, 2, \ldots, m\}$, each letter i occurring with known probability $p_i > 0$. Let S be the random codeword-length resulting from letter-by-letter encoding of the source output. It is desired to find a decipherable code that minimizes the expected value of a^S. Establish the lower bound $E(a^S) \geq (\sum_{1}^{m} \sqrt{p_i})^2$, and characterise when equality occurs.

Hint: employ the Cauchy-Schwarz inequality, that (for positive x_i, y_i)

$$\sum_{1}^{m} x_i y_i \leq \left(\sum_{1}^{m} x_i^2\right)^{\frac{1}{2}} \left(\sum_{1}^{m} y_i^2\right)^{\frac{1}{2}},$$

with equality if and only if $x_i = cy_i$ for all i.

Prove also that an optimal code for the above criterion must satisfy $E(a^S) < a(\sum_{1}^{m} \sqrt{p_i})^2$. (Cambridge 1990, and cf. Campbell (1965))

8.   For $0 < \alpha < 1$ and for $\alpha > 1$ the *information gain of order α* is defined for distributions $p = (p_i)_{i=1,\ldots,m}$ and $q = (q_i)_{i=1,\ldots,m}$ as follows:

$$I_\alpha(p|q) := \begin{cases} (\alpha - 1)^{-1} \sum_{\{i:q_i>0\}} p_i^\alpha q_i^{1-\alpha} & \text{if } p_i = 0 \text{ whenever } q_i = 0, \\ +\infty & \text{otherwise.} \end{cases}$$

Show that $I_\alpha(p|q) \geq 0$, with equality if and only if $p = q$.

The letters of a source alphabet of size m have probabilities p_1, ..., p_m, and are coded by a decipherable code using a code alphabet of size a. Let S be the random codeword-length resulting from letter-by-letter encoding of the source output. Using the I_α inequality, with suitable q and α, and the Cauchy-Schwarz and Kraft inequalities, show that

$$E(a^S) \geq 1 \Big/ \sum_i p_i^2.$$

(Cambridge 1975)

9.   An alphabet F_m of m letters, with probabilities $p_1, \ldots, p_m$, is to be encoded into an alphabet G_a of a letters where $a < m$. Denote by s_i the length of the word that encodes the ith letter of F_m.
(a) A proper subset B of $\{1, \ldots, m\}$ is given, of $k \geq 2$ elements. Prove that a decipherable coding such that

$$s_i = r \qquad (i \in B),$$

where r is the least integer with $ka^{-r} < 1$, minimizes $\max_{i \in B} s_i$ over all decipherable codings.
(b) Let F_m be indexed so that $p_1 \geq p_2 \geq \cdots \geq p_m > 0$, and suppose there exists k such that $p_1 + \cdots + p_k = \frac{1}{2}$. Let S be the random word-length of a coding $F_m \to G_a$ and define the *median* of S to be the least integer s such that $P(S \leq s) \geq \frac{1}{2}$. Prove that a coding as in (a), with $B := \{1, \ldots, k\}$, minimizes the median of S over all decipherable codings.
      Find such a coding when $m = 5$, $a = 2$ and the p_i are $\frac{1}{4}, \frac{1}{4}, \frac{1}{4}, \frac{1}{8}, \frac{1}{8}$. (Cambridge 1989)

10.   Let f be an optimal binary prefix-free code for a Bernoulli source $\mathbf{U}$ with probability distribution p. Integers n and k are fixed (and are known to the decoder) and the output of the kth extension $\mathbf{U}^{(k)}$ of the source is encoded as follows: for a source-segment $u_1 \ldots u_k$ take the concatenated string $f(u_1) \ldots f(u_k)$, of length s say. If $s \leq n$ append $n - s$ zeros to the string; if $s > n$ take the first n symbols of the string and discard the rest. This gives a 'block' code, with all codewords of length n, but at the expense of some chance of error.
      Show that given $\varepsilon > 0$ you can by taking k suitably large code in this way, with some $n \leq kh(p)$, such that the probability of error is less than ε.
      (This needs the probability theorem known as the Weak Law of Large Numbers: see §2.5.)

11.   A code is to be found such that the first symbol of a codeword is

always 0 or 1, while any later symbols can be 0, 1 or 2.
(a) Find a general method of obtaining an optimal (minimal expected codeword-length) prefix-free code. Try it out on the sources with probability distributions

$$\cdot25 \quad \cdot25 \quad \cdot2 \quad \cdot1 \quad \cdot1 \quad \cdot1,$$
$$\cdot25 \quad \cdot25 \quad \cdot25 \quad \cdot1 \quad \cdot05 \quad \cdot05 \quad \cdot05.$$

(b) Observe that among codes satisfying the constraint the code 0, 1, 02 is decipherable, though not prefix-free. Show that for certain distributions on 3-letter alphabets it is better than the constrained-optimal prefix-free code.

12.   Consider binary codewords ordered lexicographically (see Exercise 1.2.4). Suppose source letters $1, \ldots, m$ have respective probabilities $p_1 \geq \cdots \geq p_m$, all positive. Show that there exists an optimal prefix-free code f with $f(m) > f(m-1) > \cdots > f(1)$.
    Hint: first find the word-lengths s_i for a Huffman code. Then write down nodes in order for the source letters $m, m-1, \ldots, 1$ and, by always joining two nodes of lowest weight, try to build a tree with the given branch-lengths s_i. You must do it with no branches crossing! Try some examples.

13.   (a) Show that for a source with probability distribution $\frac{1}{3}, \frac{1}{3}, \frac{1}{4}, \frac{1}{12}$ both the code 0, 10, 110, 111 and the code 00, 01, 10, 11 are optimal binary encodings.
    (b) Consider the class $\mathcal{P}$ of probability distributions on a 4-element set, written as column 4-tuples $p = (p_1, p_2, p_3, p_4)^\top$, with $p_1 \geq p_2 \geq p_3 \geq p_4$. Let $\mathcal{Q}$ be the set of those $p \in \mathcal{P}$ for which both the above codes are optimal. Prove that $\mathcal{Q} = \{p \in \mathcal{P} : p_1 = p_3 + p_4\}$.
    (c) Find $q^{(1)}, q^{(2)}, q^{(3)}$ in $\mathcal{Q}$ such that $\mathcal{Q}$ is their 'convex hull': every $q \in \mathcal{Q}$ can be written $q = \lambda_1 q^{(1)} + \lambda_2 q^{(2)} + \lambda_3 q^{(3)}$ where $\lambda_i \geq 0$ and $\lambda_1 + \lambda_2 + \lambda_3 = 1$.
    Hint: making two of the inequalities $p_1 \geq p_2 \geq p_3 \geq p_4$ into equalities gives you two equations which with $p_1 + p_2 + p_3 + p_4 = 1$ and $p_1 = p_3 + p_4$ define a single element of $\mathcal{Q}$.

14.   In the Huffman tree-building algorithm suppose that when you form a new node of weight $p + p'$ out of nodes of weights p, p', you place it as *high* as possible among weights of the same value, i.e. between p_j and p_{j+1} where $p_j > p + p' \geq p_{j+1}$.
    (a) Show that this method gives the smallest value of $\sum_i s_i$ and of $\max_i s_i$ among all decipherable codes that minimize $\sum_i p_i s_i$.
    (Schwartz (1964))

(b) Show that it also gives the smallest value of $\sum p_i s_i^2$ among all decipherable codes that minimize $\sum_i p_i s_i$.

15. Consider a-ary Huffman encoding of a source with m-letter alphabet of equally likely letters. Assume that $m > a$, and also that $m \equiv 1$ (mod $a - 1$), that is, $m = 1 + k(a - 1)$ for some integer k. The algorithm is then straightforward (see Exercise 1.7.4). Let j be the integer such that $a^j \le m < a^{j+1}$, and define r by $m = a^j + r(a - 1)$. Observe that r is an integer! Show that the Huffman code consists of $a^j - r$ words of length j, and ra words of length $j + 1$.

16. A set of m apparently identical coins consists of $m - 1$ of equal weight and one that is heavier. The heavier coin is to be determined by use of a balance, on which subsets of the coins may be balanced against each other, to determine which subset (if either) contains the heavier coin. It is desired to minimize the expected number of weighings required.

Relate this problem to a coding problem in which the code alphabet has three symbols, and hence determine bounds on the minimum expected number of weighings required. Determine an optimal weighing scheme and show that it minimizes also the maximum number of weighings required. When can the lower bound on the expected number of weighings be attained? (Cambridge 1980)

PROPERTIES OF A MESSAGE SOURCE

2.1 Introduction: probability

An economical representation, for instance a Huffman code as in §1.7,

- expresses the source message in the available alphabet;
- used segmentally, compacts the source message.

The latter fact is based on the assertion that h_n/n decreases with n for sources that occur in practice, as we stated at the end of §1.5. (Recall that h_n/n is close to the minimal mean length of codewords per source letter in an economical representation.) We shall study sources to substantiate this assertion. As well as some general theory we will define and consider Markov, stationary and ergodic sources, and look again at the Bernoulli sources that we have met already. We will also look at two naturally occurring 'sources' from an informational viewpoint: English written text, and the bases of a DNA strand.

Our intention in relation to all sources is to identify 'rate of information-flow' in a suitable sense, and bounds on it under 'source coding'. The idea of the latter is that by allowing a non-zero but asymptotically negligible error-rate, compression of the data stream can be realized. We will study in this chapter to what extent this is in principle achievable, rather than specific codes to accomplish it. The source end of the transmission diagram in Chapter 0 now expands as shown.

$$\boxed{\text{Source}} \longrightarrow \boxed{\substack{\text{Economical} \\ \text{representation}}} \longrightarrow \boxed{\substack{\text{Source} \\ \text{coding}}} \longrightarrow \boxed{\substack{\text{Channel} \\ \text{coding}}} \longrightarrow \cdots$$

Two substantial tasks that the above gives rise to are to develop the theory of finite Markov chains, for use in our consideration of Markov sources, and to develop properties of the entropy function h that we met in Chapter 1. We will treat Markov chains, in §2.6, algebraically, which should make a worthwhile contrast to the analytic approach you may have met already or are likely to meet in probability courses. Entropy will be considered in detail in §§2.9–10, and will be the key to our understanding of quantitative information.

In later chapters we meet the *noisy channel*, which introduces random errors. For that we need another sort of coding: *error-correcting codes*. They may be used as well as economical representations and source codings. Our criterion for a good code will eventually shift to minimizing the chance that any errors (at all) creep through into the received and decoded message: *reliable transmission*.

Probability spaces

It is appropriate to give at this point a concise treatment of the probability set-up that underlies the idea of a random source. You should be familiar with discrete random variables. A *discrete r.v.* is simply a function from one set, the *sample space* Ω, to the real line, that takes only countably many values. So its set of values is either finite: $\{a_1, \ldots, a_n\}$, or can be indexed as an infinite sequence: $\{a_1, a_2, \ldots\}$. We are thinking here of one of the stream of random variables, say U_1, that we modelled the source by in §1.4. They take values in $\{1, \ldots, a\}$. The simplest sample space for a single such r.v. U_1 would be $\Omega := \{1, \ldots, a\}$; then U_1 could be the identity function $U_1(\omega) := \omega$ for $\omega \in \Omega$.

Assuming then, as we may, that Ω has at most countably many elements (the *sample points* $\omega \in \Omega$), there is defined on it a *probability mass function* (p.m.f.), any function $p : \Omega \to [0,1]$ with the property that $\sum_{\omega \in \Omega} p(\omega) = 1$. Subsets of Ω are called *events*, and the class of all events, here just the class of all subsets of Ω, is denoted by $\mathcal{A}$. The *probability* $P(A)$ of an event A is simply the sum of the masses assigned by p to the sample points:

$$P(A) := \sum_{\omega \in A} p(\omega) \qquad (A \in \mathcal{A}).$$

The *probability space* is the triple $(\Omega, \mathcal{A}, P)$.

The phenomenon that is being modelled by all this structure is the *repeatable experiment*. When the experiment is performed a single sample point ω is chosen by 'chance' from Ω. On repeating the experiment another, ω', is chosen, and so on. In the long run the proportion of sample points picked that have some particular value ω_0 should approximate $p(\omega_0)$.

As mentioned, a discrete r.v. is a function $X : \Omega \to \mathcal{X}$ where $\mathcal{X}$ is some finite or denumerably infinite subset of $\mathbb{R}$. For subsets B of $\mathcal{X}$ or more generally of $\mathbb{R}$ we write $P(X \in B)$ as shorthand for $P(\{\omega : X(\omega) \in B\})$, $P(X = x)$ for $P(\{\omega : X(\omega) = x\})$, and so on. The p.m.f. of X is $p_X : \mathcal{X} \to [0,1]$ defined by $p_X(x) := P(X = x)$. The *distribution* of a r.v. X is strictly the function $B \mapsto P(X \in B)$, but for X discrete we normally use the phrase 'distribution of X' loosely to mean the p.m.f.

considered as a sequence or vector $\left(p_X(x)\right)_{x \in \mathcal{X}}$.

You are assumed familiar with, and we have already used, the *expectation* or *mean* of (the distribution of) X, namely EX defined as $\sum_{x \in \mathcal{X}} x p_X(x)$ when the sum converges absolutely. For functions $g : \mathbb{R} \to \mathbb{R}$ it may be proved that $Eg(X) = \sum_{x \in \mathcal{X}} g(x) p_X(x)$ if absolutely convergent.

The random source

Of course a single r.v. is not at all enough for our purposes. We need to be able to consider the whole stream U_1, U_2, ... of such r.v.s together. A way to do that is as follows.

Suppose we have the U_n defined on distinct discrete probability spaces $(\Omega_n, \mathcal{A}_n, P_n)$. For definiteness suppose that each Ω_n is the set $\{1, \ldots, a\}$, that $\mathcal{A}_n$ is the class of all its subsets, and that each U_n is the identity map,

$$U_n(\omega_n) := \omega_n \qquad (\omega_n \in \Omega_n).$$

Now we can define a grand overall sample space Ω to be the Cartesian product of the component Ω_n:

$$\Omega := \prod_{n=1}^{\infty} \Omega_n.$$

The generic element of Ω is a sequence $\omega = (\omega_1, \omega_2, \ldots)$, where each $\omega_n \in \Omega_n$. The class $\mathcal{A}$ of events is now not *all* subsets of Ω but the smallest class containing all *cylinders*

$$A_1 \times \cdots \times A_n \times \Omega_{n+1} \times \Omega_{n+2} \times \cdots,$$

where $A_k \in \mathcal{A}_k$, that is closed under all set operations of taking countable unions or intersections, complementation and set-difference. Now a probability P defined on cylinders and hence on finite unions of cylinders has to satisfy some obvious consistency conditions, because you can express a given finite union of cylinders in terms of component cylinders in various ways. Supposing there is a probability P defined on cylinders, and satisfying these consistency conditions, a deep result known as the *Kolmogorov Extension Theorem* establishes the existence of a unique P on $\mathcal{A}$ consistent with its given values on cylinders. Each r.v. U_n is then the *projection* on the relevant axis:

$$U_n(\omega_1, \omega_2, \ldots) := \omega_n \qquad ((\omega_1, \omega_2, \ldots) \in \Omega).$$

What we have here described is the *canonical construction* for a countable collection of discrete r.v.s on a common probability space. Again, we write $P(U_1 \in A_1, U_2 \in A_2)$ for $P(\{\omega : U_1(\omega) \in A_1, U_2(\omega) \in A_2\})$, and so on. Specifying P on cylinders is equivalent to specifying all finite-dimensional joint probabilities $P(U_1 \in A_1, \ldots, U_n \in A_n)$, or, more succinctly, all probabilities $P(U_1 = u_1, \ldots, U_n = u_n)$ for $u_k \in \Omega_k$. An example of such a specification is

$$P(U_1 = u_1, \ldots, U_n = u_n) := p_1(u_1) p_2(u_2) \cdots p_n(u_n) \qquad (\text{all } n \geq 1, \ u_k \in \Omega_k),$$

where each p_k is a p.m.f. on the relevant Ω_k. This makes the U_k *independent* r.v.s. It is however not at all the only way to specify the 'joint distribution' of the U_k.

As a contrasting example, suppose that, for all n,
$$P(U_1 = U_2 = \cdots = U_n = 1) = \tfrac{1}{2} = P(U_1 = U_2 = \cdots = U_n = 2).$$
This corresponds to deciding on the toss of a fair coin that all the U_k should be 1, or that they all should be 2. They are thus far from being independent, though they all have the same individual distribution $P(U_k = 1) = \tfrac{1}{2} = P(U_k = 2)$.

We should finally mention that jointly distributed r.v.s $U_1, \ldots, U_n$ can be assembled as a 'random vector' $U^{(n)} = (U_1, \ldots, U_n)^\top$. We can extend the notion of p.m.f. to it by defining $p_{U^{(n)}}(u^{(n)}) := P(U_1 = u_1, \ldots, U_n = u_n)$ where $u = (u_1, \ldots, u_n)^\top$. This is also called the 'joint p.m.f.' of $U_1, \ldots, U_n$.

Exercise

1.   Another fundamental probability space has Ω the unit interval $[0, 1)$ and $P(A)$ simply the *length* (technically, 'measure') of A, the class $\mathcal{A}$ of events being those for which length can be defined. Taking this space for granted, define functions $U_i : \Omega \to \{0, 1\}$ by

$$U_1 := \mathbf{1}_{[\frac{1}{2},1)},$$
$$U_2 := \mathbf{1}_{[\frac{1}{4},\frac{1}{2})} + \mathbf{1}_{[\frac{3}{4},1)},$$
$$U_3 := \mathbf{1}_{[\frac{1}{8},\frac{1}{4})} + \mathbf{1}_{[\frac{3}{8},\frac{1}{2})} + \mathbf{1}_{[\frac{5}{8},\frac{3}{4})} + \mathbf{1}_{[\frac{7}{8},1)},$$
$$\cdots\cdots$$

$$\cdots\cdots$$

Show that $P(U_i = 0) = \tfrac{1}{2} = P(U_i = 1)$ for each i and that the U_i are mutually independent.

   (The U_i are the *Rademacher functions*. It is interesting to note that $U_1(\omega) = 1$ if and only if the ith digit in the binary expansion of ω is 1.)

2.2   Reliable encoding rates: the source information-rate

The source produces a stream $U_1, U_2, \ldots$ of r.v.s taking values in an alphabet $F_a = \{1, \ldots, a\}$, say. Possibly these are already-encoded words in an economical representation, possibly not.

Suppose we take the initial source output in a big segment or block $U = U^{(n)} := (U_1, \ldots, U_n)^\top$. Its generic value is $u = u^{(n)} := (u_1, \ldots, u_n)^\top$ where each $u_i \in F_a$. We shall often write this as an n-string $u_1 u_2 \ldots u_n$ (not a product!). Either way we can consider u as an element of the Cartesian product $F_a^n = F_a \times \cdots \times F_a$. Now u has a^n possible values. We define r by

$$a^n = 2^{nr}.$$

If u is sent over time n then we are transmitting at r *bits/unit time*, where $r := \log a$. So r is a 'binary equivalent rate'.

It will be enough if *most* messages can be sent at that rate, as follows:

Definition. *Suppose we can find, for each n, a set $A_n \subseteq F_a^n$ having at most r_n elements, where*

$$r_n \to r, \qquad P(U^{(n)} \in A_n) \to 1 \qquad (n \to \infty).$$

Then the source is reliably encodable at rate r.

If $U^{(n)}$ is emitted over time n then again the units of r are bits/unit time. The idea of the above is that in principle you can encode at rate almost r with negligible error for long enough blocks.

Write $\#$ for 'number of elements of'. Then the size-restriction on A_n above is that $\# A_n \leq 2^{nr_n}$. In fact we can remove the need to mention the sequence r_n at all if we write the conditions on A_n as

$$\# A_n \leq 2^{n(r+o(1))}, \qquad P(U^{(n)} \in A_n) \to 1 \qquad (n \to \infty); \qquad (1)$$

here $o(1)$ denotes some unspecified sequence of real numbers that tends to 0. We shall use this formulation quite often below.

A third way to state the size-restriction on A_n, but one that is less immediately intelligible, is

$$\limsup_{n \to \infty} n^{-1} \log(\# A_n) \leq r.$$

Definition. *The* information rate $H = H(\mathbf{U})$ *of the source $\mathbf{U}$ is the infimum of the rates r at which the source is reliably encodable.*

Roughly, nH is the minimal number of bits required to encode $U^{(n)}$. The infimum is over all choices of the sequence of sets that have the required properties.

You should be clear that H is *not* the source entropy h. Of course, h and H have something to do with each other, and we shall establish connections below. But they are conceptually distinct.

Proposition 2.2.1. *For a source with alphabet F_a,*

$$0 \leq H \leq \log a,$$

both bounds being attainable.

Proof. Exercise 1.   □

Telegraph English

This is a good point to consider what H might be for one existing 'source', namely English text with punctuation removed but spaces retained, and written in lower-case. Including the space character, the alphabet has thus $a = 27$ letters. Since $27^n \simeq 2^{4\cdot755n}$ you can code telegraph English at a rate of 4·76 bits/letter. However considerably lower rates suffice. Coding segments of length 3, for instance, about 3·1 bits/letter should be enough, and the true H for English text appears to be below 1·3 bits/letter. You will find lengthy discussions of this, with references, elsewhere (e.g. Yaglom & Yaglom (1983), §4.3), so we mention here just a few aspects.

Table 2.2.1. An economical representation for English text

Letter	Code	Letter	Code	Letter	Code	Letter	Code
␣	11	r	1001	u	011001	x	1000111
t	0101	s	01111	d	011000	g	1000110
e	0100	h	01110	f	001011	k	00101001
a	0011	c	01101	w	101011	q	001010001
i	0001	p	00100	b	101010	j	0010100001
o	0000	l	10100	y	100010	z	0010100000
n	1011	m	10000	v	0010101		

The point is that English has *redundancy*: much less content per letter is conveyed than if the letters were independently chosen. A way to see this if you need convincing is to decode by hand the following message which is in the code of Table 2.2.1:

0110000100011010000011000000110111000110110001011111101000001

1011111000101100010010111110001000000110011100101001101100001

10101111011100000101011110101000011011110010001001010010100.

You will have found that you can complete words once started on them, and even correctly guess the next word in some cases.

Some consequences of the redundancy of natural language are the following.

(i)   Data compaction is possible, as we have discussed. An extreme case of this is form letters as used by many businesses, where whole paragraphs are set up to be typed at the touch of a single key.

(ii)   Error identification and correction are possible. For instance the whole text of this book has been put through a 'spelling checker' computer program, to find misprints.

(iii)   Code-breaking is possible. For instance, it is the case that with a message of 40 letters or more one can expect to break a simple 'secret' code got by a cyclic permutation of the letters of the alphabet.

(iv)   Crossword puzzles can be solved.

Example 2.2.2. Consider the special sort of crossword with no black space, as in Fig. 2.2.2. Thickened cell-boundaries denote the divisions between words. Languages vary somewhat in their redundancy, and in a (non-existent) language with no redundancy, all sequences of letters would be grammatically correct, so that any array of this type would be a valid solution to such a puzzle. In solving it there would be nothing to go on but the clues and that each letter is part of two distinct words.

a	n	o	d	e
b	i	k	e	r
a	f	t	e	r
s	t	o	r	e
e	y	o	t	m

Figure 2.2.2

By contrast, in a language with too much redundancy such an array of any substantial size would be impossible to compile, as there would be too many restrictions on possible letter-strings.

Exercises

1.   If a source emits one character (from an alphabet of a) every unit of time, show that its information rate H bits/unit time has the bounds

$$0 \leq H \leq \log a,$$

and for each bound give an example of a source for which equality occurs.

2.   Consider a source that emits symbols $X_1, X_2, \ldots$, independent r.v.s each taking values 0, 1 with probabilities $\frac{1}{2}, \frac{1}{2}$. The mode of operation is unusual: X_1 is transmitted once, X_2 twice, X_3 three times, and so on. Thus, if X_1 is transmitted at time 0, X_n is first transmitted at time $1 + 2 + \cdots + (n - 1) = (n - 1)n/2$. Prove that the source is reliably encodable at rate 0.

3.   Information rate is defined as the infimum of reliable encoding rates. Show that the infimum is attained; that is, H is itself a reliable encoding rate.

   Hint: choose any sequence $r^{(k)} \downarrow H$ and for each k let $A_n^{(k)}$ be a sequence of sets satisfying (1) for $r := r^{(k)}$. Then build a sequence of sets A_n from successive blocks of the $A_n^{(1)}$, the $A_n^{(2)}$, and so on, in such a way that $\# A_n^{(k)} \leq 2^{r^{(k)}+1/k}$ and $P(U^{(n)} \in A_n^{(k)}) > 1 - k^{-1}$ for all those $A_n^{(k)}$ that appear in the kth block.

2.3   The First Coding Theorem (FCT)

The First Coding Theorem gives an evaluation of the information rate under a sufficient condition. This of course throws the problem back to the verification of the sufficient condition, but we will find later that that can often be done. The condition is in terms of a certain sort of convergence of random variables, so we first need to discuss what sorts there are.

Convergence of random variables
Let ξ, ξ_1, ξ_2, ... be r.v.s on a common probability space.

Definitions. *ξ_n converges* in probability *to ξ, notation $\xi_n \xrightarrow{\text{P}} \xi$, if, for all $\varepsilon > 0$,*

$$P(|\xi_n - \xi| \le \varepsilon) \to 1 \qquad (n \to \infty).$$

ξ_n converges almost surely (a.s.) to ξ, notation $\xi_n \xrightarrow{\text{a.s.}} \xi$, if

$$P(\xi_n \to \xi \text{ as } n \to \infty) = 1.$$

For r.v.s ξ_n with finite expectations we say that ξ_n converges in L^1 to ξ, notation $\xi_n \xrightarrow{L^1} \xi$, if

$$E|\xi_n - \xi| \to 0 \qquad (n \to \infty).$$

For r.v.s ξ_n with finite second moments $E(\xi_n^2)$ we say that ξ_n converges in L^2 to ξ, notation $\xi_n \xrightarrow{L^2} \xi$, if

$$E(|\xi_n - \xi|^2) \to 0 \qquad (n \to \infty).$$

It is probably best to wait to see how these types of convergence are used in the text, rather than think about them deeply at this stage. However, some remarks are worth making now.

The convergences have different strengths: so some are harder to establish, hold in fewer circumstances, and have correspondingly stronger consequences. The implications between them are as shown:

$$
\begin{array}{c}
\text{a.s.} \\
\Downarrow \\
L^2 \implies L^1 \implies \text{P} \, .
\end{array}
$$

No other implications hold in general.

A.s. convergence says that the event $\{\omega : \xi_n(\omega) \to \xi(\omega) \text{ as } n \to \infty\}$ has probability 1. (The reason why this ω-set is indeed an event is that it can be written $\bigcap_{m=1}^{\infty} \bigcup_{n=1}^{\infty} \bigcap_{k=1}^{\infty} \{\omega : |\xi_{n+k}(\omega) - \xi(\omega)| \le \frac{1}{m}\}$, so is obtainable from cylinder sets by countably many operations.) So apart from a *null* event, one of zero probability, the actual convergence of ξ_n to ξ does occur. It is pointwise convergence of functions, off a negligible set.

The other modes of convergence are all weaker in the sense that they say only that, *for each large n separately,* ξ_n is likely to be close to ξ. They do not assert any approach to ξ as you keep on observing successive ξ_n. This is most visibly so in the definition of convergence in probability, but it is the case also for the integral convergences L^1 and L^2. If $\xi_n \xrightarrow{P} \xi$ and the second moments (or variances) of the ξ_n are uniformly bounded:

$$E(\xi_n^2) \leq C < \infty \qquad (n \geq 1), \tag{1}$$

then $\xi_n \xrightarrow{L^1} \xi$. This is the way that L^1 convergence usually is established, and is the best idea of it to keep in mind. The main extra *consequence* of L^1 convergence over convergence in probability is that $E\xi$ exists, is finite, and $E\xi_n \to E\xi$.

Similarly $\xi_n \xrightarrow{L^2} \xi$ follows from $\xi_n \xrightarrow{P} \xi$ and uniformly bounded $(2+\delta)^{\text{th}}$ absolute moments: $E(|\xi_n|^{2+\delta}) \leq C < \infty$ for some $\delta > 0$. And L^2 convergence implies convergence of variances: $\operatorname{var} \xi_n \to \operatorname{var} \xi < \infty$. (See, for example, Loève (1977), §9.4 for these implications. You are asked to prove some of them in Exercises 2 and 3.)

The First Coding Theorem
The p.m.f. of $U^{(n)}$ is p_n given by

$$p_n(u) := P(U^{(n)} = u) \qquad (\text{all } u \in F_a^n).$$

Since $p_n : F_a^n \to \mathbb{R}$ and $U^{(n)} : \Omega \to F_a^n$, you can form $p_n(U^{(n)})$, that is, $\omega \mapsto p_n(U^{(n)}(\omega))$, a random variable.

Forming this peculiar composition is a characteristic feature of information theory, not found elsewhere in probability theory.

Example 2.3.1. Suppose $F_a = \{A, B, C\}$ and that

$$U^{(2)} = \begin{cases} AB & \text{with probability } \cdot3, \\ AC & \text{with probability } \cdot1, \\ BC & \text{with probability } \cdot1, \\ BA & \text{with probability } \cdot2, \\ CA & \text{with probability } \cdot25, \\ CB & \text{with probability } \cdot05. \end{cases}$$

Thus $p_2(AB) = \cdot3$, $p_2(AC) = \cdot1, \ldots, p_2(CB) = \cdot05$. So $p_2(U^{(2)})$ takes values

$$\cdot3, \text{ with probability } \cdot3,$$
$$\cdot1, \text{ with probability } \cdot2,$$
$$\cdot2, \text{ with probability } \cdot2,$$
$$\cdot25, \text{ with probability } \cdot25,$$
$$\cdot05, \text{ with probability } \cdot05.$$

Some 'lumping together' has occurred.

Now for the First Coding Theorem. Let
$$\log_+ x := \begin{cases} \log x & \text{if } x > 0 \\ 0 & \text{if } x = 0 \end{cases},$$
and define the r.v.
$$\xi_n := -n^{-1}\log_+ p_n(U^{(n)}).$$

Theorem 2.3.2: the First Coding Theorem (FCT). *If $\xi_n \xrightarrow{P} c$ for some non-random c then $H = c$.*

Proof. Choose $\varepsilon > 0$, then
$$B_n := \{u \in F_a^n : |-n^{-1}\log_+ p_n(u) - c| \le \varepsilon\}$$
has $P(U^{(n)} \in B_n) \to 1$. Remove from B_n all u with $p_n(u) = 0$; this does not alter $P(U \in B_n)$. Then $p_n(u) \ge 2^{-n(c+\varepsilon)}$ for all $u \in B_n$, so
$$P(U \in B_n) \ge 2^{-n(c+\varepsilon)} \# B_n,$$
from which it follows that $\# B_n \le 2^{n(c+\varepsilon)}$. Thus the source is reliably encodable at rate $c + \varepsilon$, whence $H \le c$.

If $c = 0$ we are done. Otherwise, pick ε satisfying $0 < \varepsilon < \frac{1}{2}c$ and consider any $A_n \subseteq F_a^n$ with $\# A_n \le 2^{n(c-2\varepsilon+o(1))}$. With B_n as above, $p_n(u) \le 2^{n(c-\varepsilon)}$ for all $u \in B_n$, so
$$P(U \in A_n \cap B_n) \le 2^{-n(c-\varepsilon)} \#(A_n \cap B_n) \le 2^{-n(c-\varepsilon)}2^{n(c-2\varepsilon+o(1))} \to 0.$$
But then
$$P(U \in A_n) \le P(U \in A_n \cap B_n) + P(U \in B_n^c) \to 0 + 0 = 0,$$
and this shows that the source is *not* reliably encodable at rate $c - 2\varepsilon$. So $H \ge c$. $\square$

Exercises

1.   Show that $x\log_+ x$ is continuous on $[0,\infty)$. (This is the point of using $\log_+$.) Sketch the function. What is its infimum and where is it attained?

2.   *Chebychev's inequality* states that, for a random variable X and any positive ε, $P(|X - EX| \ge \varepsilon) \le \varepsilon^{-2}\operatorname{var} X$. It thus gives a bound on the probability of X being far from its (finite) mean EX, in terms of its variance $\operatorname{var} X = E((X - EX)^2)$. You will find a proof in any probability or statistics textbook.
     (a) Use Chebychev's inequality to show that L^1 convergence of r.v.s implies convergence in probability, to the same limit. Find a case where convergence in probability holds but not L^1 convergence.

(b) By writing $E|X_n| = E(|X_n|\mathbf{1}_{|X_n|\leq 1}) + E(|X_n|\mathbf{1}_{|X_n|>1})$ and using the fact that $|x| \leq x^2$ for $|x| > 1$, show that $X_n \xrightarrow{L^2} 0$ implies $X_n \xrightarrow{L^1} 0$.

(The notation $\mathbf{1}$ should be self-explanatory, but is in any case formally set up in §2.11.)

(c) Deduce that L^2 convergence of r.v.s implies L^1 convergence, to the same limit.

3.   (a) Show that if $\xi_n \xrightarrow{L^1} \xi$ then $E|\xi_n| \to E|\xi| < \infty$ and $E\xi_n \to E\xi$.

(b) Show for any real numbers x, y that $(x+y)^2 \leq 2x^2 + 2y^2$. Deduce that if $\xi_n \xrightarrow{L^2} \xi$ then $E(\xi_n^2) \to E(\xi^2) < \infty$ and $\mathrm{var}\,\xi_n \to \mathrm{var}\,\xi < \infty$.

2.4   Asymptotic Equipartition Property (AEP)

If the condition of the FCT holds then $\xi_n \xrightarrow{P} H$, and this is equivalent to the assertion that, for all $\varepsilon > 0$,

$$P\big(2^{-n(H+\varepsilon)} \leq p_n(U^{(n)}) \leq 2^{-n(H-\varepsilon)}\big) \to 1 \qquad (n \to \infty).$$

More clearly, the FCT says that a source satisfying its condition has the following property:

> For all $\varepsilon > 0$ there exists $n_0(\varepsilon)$ such that for all $n \geq n_0(\varepsilon)$ the set F_a^n decomposes into disjoint sets Π_n, T_n such that
> $$P(U^{(n)} \in \Pi_n) < \varepsilon,$$
> $$2^{-n(H+\varepsilon)} \leq P(U^{(n)} = u) \leq 2^{-n(H-\varepsilon)} \qquad \text{for all } u \in T_n.$$

The framed statement is called the Asymptotic Equipartition Property (AEP). It says the possible values of $U^{(n)}$ are those in the *typical set* T_n and the *residual set* Π_n, the latter of vanishing probability. In the typical set the strings all have probabilities approximately equal to 2^{-nH}.

The moral of this is that for a source with the AEP, as the typical strings are approximately equiprobable (equidistributed), *encode them with codewords of the same length* (a *block code*). Since Π_n has vanishing probability you can encode its contents virtually anyhow.

The AEP illuminates the definition of 'reliable encoding rate'. Suppose the source emits 1 symbol (from F_a) per second. This is equivalent to log a bits/sec, since in n seconds any of $a^n = 2^{n \log a}$ strings are possible. If the AEP holds, you need bother only with $2^{n(H+o(1))}$ of these, which are virtually equally likely to occur. So the *effective* rate is $H + o(1)$ bits/sec.

We shall prove below that large classes of sources have the AEP, and we will give means to calculate their information rates. The AEP is not universal however: we will give in §2.12 an example of a source not having it.

Note. It might occur to you that with a slight strengthening of its assumption, the FCT has not too much to say. For any $f : F_a^n \to \mathbb{R}$, $Ef(U^{(n)}) = \sum_u p_n(u)f(u)$, so on putting $f(u) := -\frac{1}{n}\log_+ p_n(u)$ you get

$$E\xi_n = -\frac{1}{n}\sum_{u \in F_a^n} p_n(u)\log_+ p_n(u) = n^{-1}h(U^{(n)}).$$

(Recall that $\frac{1}{n}h(U^{(n)})$ is the per-symbol source entropy, for n-strings.) Thus if we were to strengthen the assumption of the FCT to $\xi_n \xrightarrow{L^1} c$ (non-random) then we could conclude

$$n^{-1}h(U^{(n)}) \to H \qquad (n \to \infty)$$

(see Exercise 1). However we shall show the latter directly for a large class of sources, *without* the assumption.

Exercise

1. Prove that $\xi_n \xrightarrow{L^1} c$ (non-random) implies $\frac{1}{n}h(U^{(n)}) \to H$ as $n \to \infty$.

2.5 The information rate for a Bernoulli source

We defined a Bernoulli source in §1.5: it emits $U_1, U_2, \ldots$, these being i.i.d. (independent, identically distributed) r.v.s with values in F_a.

For Bernoulli sources the basic limit theorems of probability theory have obvious relevance so we state them here, for later reference as well.

Suppose $X_1, X_2, \ldots$ are independent r.v.s with a common distribution having finite mean μ, and put $S_n := X_1 + \cdots X_n$. The *Weak Law of Large Numbers* (WLLN) states that $S_n/n \xrightarrow{P} \mu$ as $n \to \infty$. The *Strong Law of Large Numbers* (SLLN) states that under the same conditions $S_n/n \xrightarrow{\text{a.s.}} \mu$.

(So what is the point of the Weak Law? For one thing, it is easier to prove. For another, an extended Weak Law, with convergence in probability of S_n/n to some constant, can be proved to hold with slightly less restriction on the common distribution of the X_i. This is not so for the Strong Law.)

If the common distribution of the X_n has finite variance $\sigma^2 > 0$ then the *Central-limit Theorem* (CLT) says that, for every $x \in \mathbb{R}$,

$$P\left(\frac{S_n - n\mu}{\sigma\sqrt{n}} \leq x\right) \to \int_{-\infty}^{x} \frac{1}{\sqrt{2\pi}} e^{-\frac{1}{2}z^2}\, dz \qquad (n \to \infty).$$

This brings in yet another mode of convergence.

Definition. ξ_n converges in distribution to ξ, notation $\xi_n \overset{d}{\to} \xi$, if

$$P(\xi_n \le x) \to P(\xi \le x) \qquad (n \to \infty)$$

for every x at which the function $F(x) := P(\xi \le x)$ is continuous.

This is the weakest of the modes of convergence we have met, being implied by all the others. It says only that for large n the distribution of ξ_n is close to some given distribution, but little about how the sequence $\xi_1, \xi_2, \ldots$ might behave if observed. The conclusion of the CLT can be rephrased as that $(S_n - n\mu)/(\sigma\sqrt{n}) \overset{d}{\to} Z$ where Z has the standard Gaussian or normal distribution denoted $N(0, 1)$, namely that with probability density

$$\phi(z) := (2\pi)^{-\frac{1}{2}} e^{-\frac{1}{2}z^2} \qquad (z \in \mathbb{R}). \tag{1}$$

Informally, mixing notations, one can write $(S_n - n\mu)/(\sigma\sqrt{n}) \overset{d}{\to} N(0, 1)$.

Now for the Bernoulli source we evaluate the source information rate H, as defined in §2.2.

Theorem 2.5.1. *For a Bernoulli source, for all n,*

$$H = n^{-1}h(U^{(n)}) = h(U_1),$$

and the source satisfies the AEP.

Proof. $p_n(u) = p(u_1) \cdots p(u_n)$ where u is the n-string $u_1 u_2 \ldots u_n$, and $p(\cdot)$ is the common p.m.f. of the U_i. So

$$\xi_n = -\frac{1}{n} \sum_1^n \log_+ p(U_i),$$

and the r.v.s $-\log_+ p(U_i)$ are i.i.d. with mean $h(U_1)$. Then

$$\xi_n \overset{P}{\to} E(-\log_+ p(U_1)) = h(U_1)$$

by the WLLN. Then the FCT gives the conclusion. □

Exercises

1. Consider a message source in which characters are independently distributed, U_t taking equiprobably all values in an alphabet of size a_t. Show that one can so choose a_t as a function of t that $h(U^{(n)})/n$ does not have a limit as $n \to \infty$.

2. Use Chebychev's inequality (see Exercise 2.3.2) to prove the WLLN under the extra condition that the common variance of the X_i is finite.

3. The (cumulative) *distribution function* of an r.v. X is $F : \mathbb{R} \to [0, 1]$ defined by $F(x) := P(X \le x)$ for $x \in \mathbb{R}$. Recalling what you know

about general, not necessarily discrete, r.v.s, show that the distribution function F of any r.v. X is right-continuous $(\lim_{x \downarrow x_0} F(x) = F(x_0))$ and has left limits $(\lim_{x \uparrow x_0} F(x) = P(X < x_0))$; thus $P(X = x_0) = F(x_0) - \lim_{x \uparrow x_0} F(x)$. Show that there can be at most n places where $F(x_0) - \lim_{x \uparrow x_0} F(x) \geq \frac{1}{n}$, and deduce that the set D_F of x where F is discontinuous is countable (including finite).

Thus F is determined by its values on D_F^c, and the convergence $\xi_n \xrightarrow{\mathrm{d}} \xi$ determines the distribution of ξ.

2.6 Finite Markov chains

Suppose $U_1, U_2, \ldots$ form a time-homogeneous Markov chain on F_a, that is, they satisfy the *Markov property:*

$$P(U_{n+1} = k | U_1 = u_1, \ldots, U_n = u_n) = P(U_{n+1} = k | U_n = u_n),$$

and have *stationary transition probabilities:*

$$P(U_{n+1} = k | U_n = j) = p_{jk},$$

where j, $k \in F_a$ and $n = 1, 2, \ldots$. The Markov property says that the conditional distribution of the next symbol to be emitted by the source, given all its output so far, is the same as the conditional distribution given just the latest symbol emitted. So the dependence is only 1-step, and knowledge of more than one step back is no advantage, in predicting subsequent symbols, over knowledge of just one step back.

The property of having stationary transition probabilities is just the natural property of time-homogeneity of the distributional structure. It enables us to represent the one-step conditional distributions by the *transition matrix* $P = (p_{jk})$. By a *finite Markov chain* we shall always mean one with stationary transition probabilities, on some finite set which we might as well take to be $F_a = \{1, \ldots, a\}$. The elements $1, \ldots, a$ are called its *states*. The *initial distribution* is that of U_1. The initial distribution and the transition probabilities determine the joint distribution of $U_1, \ldots, U_n$ for any n:

$$P(U_1 = u_1, \ldots, U_n = u_n) = P(U_1 = u_1) p_{u_1 u_2} \cdots p_{u_{n-1} u_n}.$$

We shall develop the theory of such chains in order to evaluate the information rate. Those proofs below that are not very information-theoretic are set in smaller type, and could be omitted on a first reading.

Lemma 2.6.1. *For any integers* $1 \leq i_1 < \cdots < i_m < n$ *and* $r > 0$,

$$P(U_{n+r} = k | U_{i_1} = u_{i_1}, \ldots, U_{i_m} = u_{i_m}, U_n = j) = P(U_{n+r} = k | U_n = j), \quad (1)$$

and the r-step transition probabilities $p_{jk}^{(r)} := P(U_{n+r} = k | U_n = j)$ have matrix $(p_{jk}^{(r)}) = P^r$, the rth power of P.

Proof. Proving (1) is easiest notationally if there are no gaps in what is given about the states of the chain up to time n, so let us consider

$$P(U_{n+r} = u_{n+r} | U_1 = u_1, \ldots, U_n = u_n) = \frac{P(U_1 = u_1, \ldots, U_n = u_n, U_{n+r} = u_{n+r})}{P(U_1 = u_1, \ldots, U_n = u_n)}$$

$$= \frac{\sum_{u_{n+1}, \ldots, u_{n+r-1}} P(U_1 = u_1, \ldots, U_n = u_n, U_{n+1} = u_{n+1}, \ldots, U_{n+r} = u_{n+r})}{P(U_1 = u_1, \ldots, U_n = u_n)}$$

$$= \frac{\sum_{u_{n+1}, \ldots, u_{n+r-1}} p_{u_1 u_2} \cdots p_{u_{n-1} u_n} p_{u_n u_{n+1}} \cdots p_{u_{n+r-1} u_{n+r}}}{p_{u_1 u_2} \cdots p_{u_{n-1} u_n}}$$

(by the Markov and stationary-transition properties)

$$= \sum_{u_{n+1}, \ldots, u_{n+r-1}} p_{u_n u_{n+1}} \cdots p_{u_{n+r-1} u_{n+r}}. \tag{2}$$

If you expand $P(U_{n+r} = u_{n+r} | U_n = u_n)$ in exactly the same way, it evaluates to the same answer. So we have proved

$$P(U_{n+r} = u_{n+r} | U_1 = u_1, \ldots, U_n = u_n) = P(U_{n+r} = u_{n+r} | U_n = u_n). \tag{3}$$

Now to deduce (1) you just need to change u_n and u_{n+r} to j and k, and to 'sum out' those U_i not required in the conditioning on the left: use the fact that

$$P(X = x | Y = y) = \sum_z P(X = x | Y = y, Z = z) P(Z = z).$$

Performing this summing-out as appropriate makes the left-hand side of (3) into the left side of (1), and leaves the right of (3) unchanged, hence (1).

We also have, rearranging (2),

$$P(U_{n+r} = u_{n+r} | U_n = u_n)$$

$$= \sum_{u_{n+r-1}} \left(\sum_{u_{n+1}, \ldots, u_{n+r-2}} p_{u_n u_{n+1}} \cdots p_{u_{n+r-2} u_{n+r-1}} \right) p_{u_{n+r-1} u_{n+r}}$$

$$= \sum_{u_{n+r-1}} P(U_{n+r-1} = u_{n+r-1} | U_n = u_n) p_{u_{n+r-1} u_{n+r}}.$$

That is,

$$p_{ik}^{(r)} = \sum_j p_{ij}^{(r-1)} p_{jk}.$$

But this says the matrices $P^{(r)} = (p_{jk}^{(r)})$ are related to one another and to $P^{(1)} = P$ by matrix multiplication: $P^{(r)} = P^{(r-1)} P$. From this it follows by induction that $P^{(r)} = P^r$. □

Geometric ergodicity

Definition. *A p.m.f. $\pi = (\pi_i)$ on F_a is called a stationary distribution for* the Markov chain (and for the transition matrix P) if

$$\sum_{j=1}^{a} \pi_j p_{jk} = \pi_k \qquad (k = 1, \ldots, a).$$

In matrix terms, with $\pi^\top$ denoting the row-vector $(\pi_1, \ldots, \pi_a)$, this says $\pi^\top P = \pi^\top$. It turns out that there always exists a stationary distribution, which furthermore is unique if the next property holds.

Definition. *The chain is* irreducible *if for every pair of distinct states j, k there exists $r \in \mathbb{N}$ such that $p_{jk}^{(r)} > 0$.*

For a Markov chain to represent output from an information source irreducibility is an entirely natural assumption: it means that any possible pattern of output must be able to recur again and again.

If π is a stationary distribution then by induction $\pi^\top P^n = \pi^\top$ for all n. So if π is actually the p.m.f. of U_m for some m:

$$P(U_m = k) = \pi_k \qquad (k = 1, \ldots, a),$$

then

$$P(U_{m+n} = k) = \sum_{j=1}^{a} P(U_m = j)P(U_{m+n} = k | U_m = j)$$

$$= \sum_{j=1}^{a} \pi_j p_{jk}^{(n)} = \pi_k,$$

that is, π is the p.m.f. of each individual U_{m+n} for $n = 1, 2, \ldots$. Hence the name 'stationary'. An alternative is 'equilibrium distribution', and that suggests more, that the distribution of U_n should 'settle down' to its equilibrium distribution as n grows large. We will prove this is so under one more condition.

Definition. *The state $j \in F_a$ is* aperiodic *if there exist coprime integers r, r' such that $p_{jj}^{(r)} > 0$ and $p_{jj}^{(r')} > 0$.*

Example 2.6.2. Suppose

$$P = \begin{pmatrix} 0 & 1 & 0 \\ \frac{1}{2} & 0 & \frac{1}{2} \\ 1 & 0 & 0 \end{pmatrix}.$$

So from state 1 the chain moves to 2, then back to 1 either directly, or via 3. For any states j, k it can go from j to k in 1, 2, or 3 steps (where '3 steps' is needed in case $j = k$). It can go from 1 back to 1 in 2 or 3 steps, or more, so although $p_{11} = 0$ you have $p_{11}^{(2)} > 0$ and $p_{11}^{(3)} > 0$, whence 1 is aperiodic since 2 and 3 are coprime (have greatest common divisor 1). In fact all three states are aperiodic, as follows at once from the following result.

Lemma 2.6.3. *For an irreducible finite Markov chain the following are equivalent:*

(i)   *there exists an aperiodic state;*
(ii)   *the chain is aperiodic, i.e. all its states are;*
(iii)   *there exists n_0 such that for all $n \geq n_0$ the matrix P^n is positive (has all entries > 0).*

For instance in the above example P^5 is a positive matrix though the lower powers are not.

Proof. Suppose i is an aperiodic state, so there are coprime r and r' such that $p_{ii}^{(r)}$ and $p_{ii}^{(r')}$ are positive. Because r and r' are coprime every sufficiently large integer m can be expressed as $m = kr + lr'$ for some non-negative* integers k and l. But then for such m we have $p_{ii}^{(m)} \geq \left(p_{ii}^{(r)}\right)^k \left(p_{ii}^{(r')}\right)^l > 0$, or, in words, the chain can go from i back to i in m steps by making k excursions each of r steps and l excursions each of r' steps. Thus $p_{ii}^{(m)} > 0$ for all sufficiently large m.

Now let j be any other state. Irreducibility means that $p_{ij}^{(n)} > 0$ and $p_{ji}^{(n')} > 0$ for some n and n'. For m as above $p_{jj}^{(n'+m+n)} \geq p_{ji}^{(n')} p_{ii}^{(m)} p_{ij}^{(n)} > 0$, so $p_{jj}^{(m)} > 0$ for all large m too. As there are only a finite number of states we can find n_0 so that $p_{jk}^{(n)} > 0$ for all $n \geq n_0$ and all states j, k.

We have thus proved that (i) implies (iii). But (iii) implies (ii) immediately, and (ii) implies (i) *a fortiori*, so the lemma is proved.   □

Now the main theorem for finite Markov chains is as follows.

Theorem 2.6.4. *An irreducible aperiodic finite Markov chain has a unique stationary distribution $\pi = (\pi_k)$, and $p_{jk}^{(n)} \to \pi_k$ as $n \to \infty$, for each j, k; indeed there exist constants $C > 0$ and $0 < \lambda < 1$ such that*

$$|p_{jk}^{(n)} - \pi_k| \leq C\lambda^n \qquad (\text{all } n \geq 1 \text{ and } j, k \in F_a). \qquad (4)$$

Proof. First, for any column a-tuple $x = (x_i)$, pre-multiplication by P is an averaging operation in the sense that

$$\min_i x_i \leq \min_i(Px)_i \leq \max_i(Px)_i \leq \max_i x_i. \qquad (5)$$

To see this, let $m = \max_i x_i$, then

$$(Px)_i = \sum_{j \in F_a} p_{ij}x_j \leq m \sum_{j \in F_a} p_{ij} = m,$$

---

*   Since r and r' are coprime *every* integer m can be expressed as $kr + lr'$ for possibly negative k and l. To show that for all $m \geq m_0$ you can insist that k and l be non-negative requires some thought: see Exercise 7.3.3. In fact the least such m_0 is $rr' - r - r' + 1$: see Pólya & Szegö (1972), Pt. 1, problem 27.

whence the right-hand inequality. You can show the left-hand inequality similarly.

If you let x be e_j, the column with 1 as its jth entry and 0 elsewhere, you find $P^n e_j$ is the jth column of P^n. Let m_n and M_n be the least and greatest elements in this column, then from (5), taking x to be $P^n e_j$, you find

$$m_n \leq m_{n+1} \leq M_{n+1} \leq M_n.$$

So both sequences m_n and M_n converge, but we need to show that the difference $d_n = M_n - m_n$ converges to 0.

Lemma 2.6.3 gives us some P^{n_0} with all entries positive. Let ε be the value of its smallest entry. We can refine (5) as follows: let m and M denote the least and greatest components of x, then $x_l = m$ for some l, and

$$
\begin{aligned}
(P^{n_0} x)_i &= \sum_{j \in F_a} p_{ij}^{(n_0)} x_j \\
&= p_{il}^{(n_0)} m + \sum_{\{j : j \neq l\}} p_{ij}^{(n_0)} x_j \\
&\leq p_{il}^{(n_0)} m + \sum_{j \neq l} p_{ij}^{(n_0)} M \\
&= p_{il}^{(n_0)} (m - M) + M \\
&\leq \varepsilon(m - M) + M.
\end{aligned}
$$

So the greatest element of $P^{n_0} x$ is bounded above by $(1 - \varepsilon)M + \varepsilon m$. Applying this fact to $-x$ you find that the least element of $P^{n_0} x$ is bounded below by $-\big((1 - \varepsilon)(-m) + \varepsilon(-M)\big) = \varepsilon M + (1 - \varepsilon)m$. The distance between the bounds is $(1 - 2\varepsilon)(M - m)$. Thus

$$\max_i (P^{n_0} x)_i - \min_i (P^{n_0} x)_i \leq (1 - 2\varepsilon)(\max_i x_i - \min_i x_i).$$

Now take x to be $P^{k n_0} e_j$, then this says

$$d_{(k+1)n_0} \leq (1 - 2\varepsilon) d_{k n_0}.$$

So the subsequence $d_{k n_0}$ converges to 0, taking the whole sequence d_n with it as the latter is monotone. Thus M_n and m_n converge to a common limit, π_j say. Then

$$
\begin{aligned}
|p_{ij}^{(n)} - \pi_j| &\leq d_n \\
&\leq d_{n_0 \lfloor n/n_0 \rfloor}
\end{aligned}
$$

(since d_n non-increasing; here $\lfloor \; \rfloor$ is integer part)

$$
\begin{aligned}
&\leq (1 - 2\varepsilon)^{\lfloor n/n_0 \rfloor} d_1 \\
&\leq (1 - 2\varepsilon)^{n/n_0} d_1 / (1 - 2\varepsilon),
\end{aligned}
$$

which proves (4) with $\lambda := (1 - 2\varepsilon)^{1/n_0}$.

Finally, the limits π_j are non-negative, and sum to 1 as is seen by letting $n \to \infty$ in the *finite* sum $\sum_{j \in F_a} p_{ij}^{(n)} = 1$. So π is a probability distribution. On letting $n \to \infty$ in $p_{ij}^{(n+1)} = \sum_{i \in F_a} p_{1i}^{(n)} p_{ij}$ you find $\pi_j = \sum_i \pi_i p_{ij}$ so π is stationary. To see that π is unique, let ρ be any stationary distribution, then $\rho^\top = \rho^\top P^n$ for all n, that is, $\rho_j = \sum_{i \in F_a} \rho_i p_{ij}^{(n)}$. Let $n \to \infty$, then you see from this that $\rho_j = \sum_{i \in F_a} \rho_i \pi_j = \pi_j$, so $\rho = \pi$ as required. $\qquad \square$

(4) says that the Markov chain is *geometrically ergodic:* the approach of $p_{ij}^{(n)}$ to π_j is governed by a geometric progression $C\lambda^n$.

The above theorem can be presented in a more general context, which identifies the value of λ in the geometric ergodicity statement and so is worth a short description. Observe that a column of 1s is a right eigenvector of P, for eigenvalue 1, as each of P's rows sums to 1. To this eigenvalue corresponds a left row eigenvector π^T say, so $\pi^\mathsf{T} P = \pi^\mathsf{T}$. As part of the *Perron-Frobenius* theory of positive matrices it can be shown that no eigenvalue of P can have modulus exceeding 1, and that there is a left eigenvector π for eigenvalue 1 with all elements non-negative. So on dividing each π_j by $\sum_j \pi_j$ one makes π a probability distribution, which is thus stationary. If P is irreducible it is the unique stationary distribution, and if P is also aperiodic then all other (complex) eigenvalues of P have modulus less than 1. Then the exact best λ in Theorem 2.6.4 is the second largest eigenvalue-modulus of P.

Exercises

1.   Verify that the transition matrix

$$\begin{pmatrix} 0 & 1 & 0 & 0 \\ 0 & 0 & 1 & 0 \\ \frac{1}{2} & 0 & 0 & \frac{1}{2} \\ 1 & 0 & 0 & 0 \end{pmatrix}$$

is irreducible and aperiodic, and that P^{10} has all entries positive but P^9 does not. Deduce that P^n has all entries positive precisely when $n \geq 10$.

2.   The transition matrix

$$P = \begin{pmatrix} \frac{1}{12} & \frac{1}{2} & \frac{5}{12} \\ \frac{1}{2} & 0 & \frac{1}{2} \\ \frac{5}{12} & \frac{1}{2} & \frac{1}{12} \end{pmatrix}$$

is symmetric so may be factorised as $P = A^\mathsf{T} \Lambda A$, where A is an orthogonal matrix whose columns are right eigenvectors of P, and Λ is a diagonal matrix with the eigenvalues of P as its diagonal. Find this factorisation and deduce explicit expressions for the entries of P^n. Verify that the rate of geometric ergodicity is governed by the second largest eigenvalue-modulus of P, which is $\frac{1}{2}$.

2.7 Markov sources

Simple Markov sources

As we have seen, a Markov chain on a state-space S is a sequence (X_n) of S-valued r.v.s satisfying the Markov property. If you 'lump together' some of the states, by considering instead $Y_n := f(X_n)$ where f is some non-injective function on S, then in general you destroy the Markov property (Exercise 1). It is an unfortunate fact that in the communication-theory literature the term 'Markov source' includes such 'hidden' Markov chains, which in general are not Markov chains themselves. So we have to speak of 'simple' Markov sources:

Definition. *A simple Markov source* **U** *is a Markov chain* (U_n) *on a finite state-space* S, *having stationary transition probabilities* (p_{ij}).

Theorem 2.7.1. *For an irreducible simple Markov source, with stationary distribution* π,

$$H = -\sum_{j}\sum_{k} \pi_j p_{jk} \log_+ p_{jk} \qquad \text{bits/unit time,}$$

and the source satisfies the AEP.

Proof (Aperiodic case only). With $\xi_n := -\frac{1}{n}\log_+ p_n(U^{(n)})$ as before, we prove

$$E\xi_n \to c := -\sum_{j}\sum_{k} \pi_j p_{jk} \log_+ p_{jk}, \tag{1}$$

$$\operatorname{var}\xi_n \to 0. \tag{2}$$

These suffice to show $\xi_n \overset{L^2}{\to} c$ because

$$E\left|\xi_n - c\right|^2 = E(\xi_n - E\xi_n)^2 + (E\xi_n - c)^2$$
$$= \operatorname{var}\xi_n + (E\xi_n - c)^2 \to 0 + 0.$$

And $\xi_n \overset{L^2}{\to} c$ implies $\xi_n \overset{P}{\to} c$, as noted in §2.3. Indeed, this is immediate by Chebychev's inequality (see Exercise 2.3.2):

$$P(\left|\xi_n - c\right| \geq \varepsilon) \leq \varepsilon^{-2} E\left|\xi_n - c\right|^2 \to 0.$$

From here the result will follow by the FCT.

To prove (1) and (2) note that, with w denoting the initial distribution,

$$p_n(u_1, \ldots, u_n) = w_{u_1} p_{u_1 u_2} p_{u_2 u_3} \cdots p_{u_{n-1} u_n},$$

so that $\xi_n = \frac{1}{n}\sum_1^n \zeta_t$ where

$$\zeta_1 := -\log w_{U_1}, \qquad \zeta_t := -\log p_{U_{t-1}U_t} \quad (t \geq 2).$$

Now

$$P(U_{t-1} = j, U_t = k) = \sum_{i \in F_a} w_i p_{ij}^{(t-2)} p_{jk}$$

$$= \sum_i w_i(\pi_j + O(\lambda^t)) p_{jk} = \pi_j p_{jk} + O(\lambda^t).$$

By $O(\lambda^t)$ we mean some quantity whose modulus is bounded by $C\lambda^t$, for some $C < \infty$, for all j, k and t. So, for $t > 1$,

$$E\zeta_t = \sum_j \sum_k P(U_{t-1} = j, U_t = k)(-\log p_{jk}) = c + O(\lambda^t)$$

whence $E(\zeta_t) \to c$ as $\to \infty$. Since convergence of a sequence implies convergence of its Cesàro averages (see Exercise 2), this yields

$$E\xi_n = (E\zeta_1 + \cdots + E\zeta_n)/n \to c.$$

For (2), similarly, for $t > 1$ and $r > 1$,

$$P(U_{t-1} = j, U_t = k, U_{t+r-1} = j', U_{t+r} = k') = \sum_i w_i p_{ij}^{(t-2)} p_{jk} p_{kj'}^{(r-1)} p_{j'k'}$$

$$= \pi_j p_{jk} \pi_{j'} p_{j'k'} + O(\lambda^t) + O(\lambda^r),$$

so

$$E(\zeta_t \zeta_{t+r}) = c^2 + O(\lambda^t) + O(\lambda^r).$$

For $t = 1$ or $r = 1$ this remains true, that is, $E(\zeta_1 \zeta_{1+r})$ is bounded in r, and $E(\zeta_t^2)$, $E(\zeta_t \zeta_{t+1})$ are bounded in t. Hence, for some C, $|\mathrm{cov}(\zeta_t, \zeta_{t+r})| \le C(\lambda^t + \lambda^r)$ for all $t, r \ge 0$. But then

$$\mathrm{var}\, \xi_n = \frac{1}{n^2} \sum_{t_1=1}^n \sum_{t_2=1}^n \mathrm{cov}(\zeta_{t_1}, \zeta_{t_2})$$

$$= \frac{1}{n^2} \left(\sum_1^n \mathrm{var}\, \zeta_t + 2 \sum_{t=1}^{n-1} \sum_{r=1}^{n-t} \mathrm{cov}(\zeta_t, \zeta_{t+r}) \right)$$

$$\le \frac{1}{n^2} \left(2nC + 2C \sum_{t=1}^{n-1} (n-t)\lambda^t + 2C \sum_{r=1}^{n-1} (n-r)\lambda^r \right)$$

$$= o(1) + \frac{4C}{n} \sum_{t=1}^{n-1} \frac{n-t}{n} \lambda^t$$

$$\le o(1) + \frac{4C}{n} \sum_{t=1}^{\infty} \lambda^t = o(1) + \frac{4C}{n(1-\lambda)} \to 0.$$

This establishes (2), hence the result.   □

Corollary 2.7.2.

$$H \le -\sum_j \pi_j \log_+ \pi_j,$$

with equality iff U_1, U_2, ... are independent. That is, the information rate of a simple Markov source is at most that of the corresponding Bernoulli source with the chain's stationary distribution as its letter-distribution.

Proof. Exercise 4 (which is a little more general).   □

Higher-order Markov sources

A Markov chain encapsulates one-step dependence of a stream of r.v.s, and a *second-order Markov chain* similarly models two-step dependence, by assuming instead of the Markov property the weaker

$$P(U_{n+1} = u_{n+1}|U_1 = u_1, \ldots, U_n = u_n)$$
$$= P(U_{n+1} = u_{n+1}|U_{n-1} = u_{n-1}, U_n = u_n).$$

Similarly for a *third-order Markov chain*, and so on.

'Telegraph English' as an information source exhibits short-term dependence between letters, associated with the redundancy we noted in §2.2, and it turns out to be approximated reasonably well by a third-order Markov chain. This is an approximation to its probabilistic behaviour, not its meaning. Higher-order Markov chains are *more predictable*, so each new U_n adds less information, and the entropy is less than for the corresponding simple-Markov and Bernoulli sources. We noted in §2.2 the decrease of per-letter entropy of English text consequent on coding in blocks, which in effect allows Markov or higher-order-Markov dependence.

Exercises

1.   Let (X_n) be a Markov chain on states 1, 2, 3, with transition matrix
$$\begin{pmatrix} 0 & \frac{1}{2} & \frac{1}{2} \\ \frac{1}{2} & \frac{1}{2} & 0 \\ 1 & 0 & 0 \end{pmatrix}$$
and with $P(X_n = i) = \frac{1}{3}$ for each i, this being the stationary distribution. Let $Y_n := f(X_n)$ where $f(1) = a$, $f(2) = b = f(3)$ and a, b are distinct symbols. Show that (Y_n) is not Markov.
      Hint: Consider, e.g., $P(Y_n = a|Y_{n-1} = b = Y_{n-2})$ and $P(Y_n = a|Y_{n-1} = b)$.

2.   Prove that convergence of a sequence of real numbers implies Cesàro convergence: $x_n \to x$ implies $\frac{1}{n}(x_1 + \cdots + x_n) \to x$.

3.   Calculate the information rate for the following Markov sources:
      (a) the general two-state chain;
      (b) a chain on states 0, 1, ..., $m-1$, where $m \geq 3$. From i the chain moves to $i+1$ (mod m) with probability p, or to $i-1$ (mod m) with probability $1-p$. What about the case $m = 2$?

4.   One would expect the inequality

$$-\sum w_j \log w_j \geq -\sum \sum w_j p_{jk} \log p_{jk}$$

where $\sum_k w_k = 1$, $\sum_k p_{jk} = 1$ for all j and $\sum_j w_j p_{jk} = w_k$. Why?
        Prove it, using the Gibbs inequality (Lemma 1.4.1). Characterise
when equality occurs.

2.8 The genetic code

Genetic information is stored in the DNA (deoxyribonucleic acid) strand,
which is effectively a sequence of bases or nucleotides, of length varying from
about 10^4 (bacteria) to 10^9 (mammals). Actually the DNA strands in each
living cell occur in twos, with pairs of complementary such strands being
wound around each other and joined by weak chemical bonds to form the
famous *double helix* of Watson and Crick. Each base in a strand can take
one of four values. With length about 10^9, there are 4^{10^9} possible different
such strands, or roughly $10^{600,000,000}$.

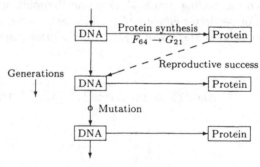

Figure 2.8.1

The DNA is directly replicated when a cell splits, thus transmitting the
information down the line of generations. The *use* of the DNA's information, and the significant coding step, occurs when proteins are synthesized
according to the pattern provided by the DNA, thus directing the development of the cell. Fig. 2.8.1 shows the process schematically: the message
in the DNA strand in one generation is transcribed to that in the next
generation, very occasionally corrupted by mutation. The existence of each
successive generation depends on the reproductive success of the organism,
built ultimately from protein synthesis directed by the DNA.
    In protein synthesis a segment of the DNA chain is first transcribed base-
by-base to a chain of messenger ribonucleic acid (mRNA) in which the four

values of the base are represented by U, C, A, G (uracil, cytosine, adenine, guanine). The protein chains are sequences of amino acids, of which there are 20 types, listed below with their conventional 3-letter abbreviations.

Alanine (ala)	Arginine (arg)	Asparagine (asn)	Aspartic acid (asp)
Cysteine (cys)	Glutamic acid (glu)	Glutamine (gln)	Glycine (gly)
Histidine (his)	Isoleucine (ile)	Leucine (leu)	Lysine (lys)
Methionine (met)	Phenylalanine (phe)	Proline (pro)	Serine (ser)
Threonine (thr)	Tryptophan (trp)	Tyrosine (tyr)	Valine (val)

For synthesis there have to be 20 instructions of the form 'append amino acid x to the protein chain', and an instruction to cease building the chain and release it, the so-called STOP element. (The START instruction is somewhat different.) There thus has to be a coding from an alphabet of 4 to one of 21, which is done by associating each instruction with a separate block of three bases on the mRNA chain. Taking the bases in threes gives an effective source-alphabet F_{64} of $4^3 = 64$ symbols, called *codons*. This is encoded into the amino-acid alphabet G_{21} by the *genetic code*, given in Table 2.8.2. With minor exceptions, it is universal among all living organisms.

Table 2.8.2. The genetic code

Codon	Amino acid	Codon	Amino acid	Codon	Amino acid	Codon	Amino acid
UUU	phe	UCU	ser	UAU	tyr	UGU	cys
UUC	phe	UCC	ser	UAC	tyr	UGC	cys
UUA	leu	UCA	ser	UAA	STOP	UGA	STOP
UUG	leu	UCG	ser	UAG	STOP	UGG	trp
CUU	leu	CCU	pro	CAU	his	CGU	arg
CUC	leu	CCC	pro	CAC	his	CGC	arg
CUA	leu	CCA	pro	CAA	gln	CGA	arg
CUG	leu	CCG	pro	CAG	gln	CGG	arg
AUU	ile	ACU	thr	AAU	asn	AGU	ser
AUC	ile	ACC	thr	AAC	asn	AGC	ser
AUA	ile	ACA	thr	AAA	lys	AGA	arg
AUG	met	ACG	thr	AAG	lys	AGG	arg
GUU	val	GCU	ala	GAU	asp	GGU	gly
GUC	val	GCC	ala	GAC	asp	GGC	gly
GUA	val	GCA	ala	GAA	glu	GGA	gly
GUG	val	GCG	ala	GAG	glu	GGG	gly

The sequence of codons (*not* the sequence of source letters U, C, A, G) can be modelled as a Bernoulli or Markov sequence. However, fairly recent research has shown that long-term repetitive patterns and shorter multi-codon switching sequences occur, not in conformity with these simple models.

Location of redundancy

It is evident from Table 2.8.2 that the necessary redundancy in coding from the 64 codons to the 21 code-symbols is located almost entirely in the last letter of the codon triplet. Altering it from C to U or vice versa always leaves the code unaltered, while altering the last letter from G to A or vice versa alters the code in only 4 out of 32 cases. This particular location of the redundancy, in the base at the *3' terminal position* in the mRNA codon, is reflected in the transcription mechanism. Transcription takes place in a ribosome, where the mRNA codon pairs with the appropriate type of tRNA (transcription RNA) molecule that has the exact *anticodon* to the mRNA codon. The 3' terminal base in the mRNA codon pairs with the base at the *5' terminal position* in the tRNA. The tRNA molecule is already carrying the appropriate amino acid, which it then parks at that location.

Now the anticodon has bases complementary to the codon: A in the codon pairs with U in the anticodon and vice-versa, and similarly G with C (so-called *Watson-Crick* base-pairing). However, according to the later *wobble hypothesis* of Crick, now confirmed, the pairing is less rigid for the third letter of each triplet: the base at the 5' terminal position in the tRNA molecule can pair in some cases with one of several possible 3' terminal bases in the mRNA molecule, as in Table 2.8.3. (The symbol I is inosine, which a group involving adenine is usually converted to at the 5' position.) Under strict Watson-Crick base-pairing 61 types of tRNA would be needed, one for every mRNA codon apart from the three that indicate STOP. Under the wobble hypothesis 32 types would suffice. It is indeed found that cells do not have the full complement of 61 types, but in practice they do have more than 32, typically 50 to 60, the composition changing with development.

Table 2.8.3

5' base of anticodon	3' base of codon
C	G
A	U
U	A or G
G	C or U
I	U, C or A

For further reading consult e.g. Zubay (1988).

2.9 Entropy

Recall that for a discrete r.v. X taking values in a finite set $\mathcal{X}$, with p.m.f. $p(x) = P(X = x)$, the *entropy* is defined to be

$$h(X) := - \sum_{x \in \mathcal{X}} p(x) \log_+ p(x) = -E \log_+ p(X)$$

where $\log_+ \alpha$ is $\log \alpha$ if $\alpha > 0$, and 0 if $\alpha = 0$, and the log is always to base 2. This makes the units of h bits. Think of $h(X)$ as the expected number of bits needed to specify X. An immediate property is

$$h(X) \geq 0, \text{ with equality iff } X \text{ is degenerate.} \qquad (1)$$

From Lemma 1.4.1, for $q(\cdot)$ a p.m.f. on $\mathcal{X}$,

$$h(X) \leq -\sum_x p(x) \log q(x), \text{ with equality iff } p = q.$$

Hence, recalling that $\# \mathcal{X}$ means the number of elements of the set $\mathcal{X}$, you obtain

Proposition 2.9.1. *If $\# \mathcal{X} = m$,*

$$h(X) \leq \log m, \text{ with equality iff } X \text{ is equidistributed over } \mathcal{X}. \qquad (2)$$

By writing down (1) and (2) for a two-point distribution $\left(\frac{p_1}{p_1+p_2}, \frac{p_2}{p_1+p_2}\right)^\top$ you deduce the *pooling inequalities*

$$-(p_1+p_2)\log(p_1+p_2) \leq -(p_1 \log p_1 + p_2 \log p_2) \leq -(p_1+p_2)\log\left(\tfrac{1}{2}(p_1+p_2)\right),$$

with equality iff $p_1 p_2 = 0$ or $p_1 = p_2$ respectively.

If ϕ is a function from $\mathcal{X}$ to some other space the distribution of $\phi(X)$ can be obtained by repeated pooling of the probabilities in the distribution of X. Any non-trivial pooling reduces the entropy. Hence

Theorem 2.9.2.

$$h(X) \geq h(\phi(X)),$$

with equality iff $X \mapsto \phi(X)$ is invertible with probability 1.

The condition means that there exists f such that $P(X = f(\phi(X))) = 1$.

Sometimes one regards $h(X)$ as a function

$$h(p_1, \ldots, p_m) = -\sum_j p_j \log_+ p_j$$

of the actual distribution $p = (p_1, \ldots, p_m)^\top$. As is easy to see,

$$
\begin{aligned}
h(p_1, \ldots, p_m) \\
= h(p_1, 1 - p_1) + (1 - p_1)h\left(\frac{p_2}{1 - p_1}, \frac{p_3}{1 - p_1}, \ldots, \frac{p_m}{1 - p_1}\right). \qquad (3)
\end{aligned}
$$

You can interpret this by thinking how to simulate X, with distribution p, by Bernoulli trials. Carry out a Bernoulli trial with probability of success p_1; if it succeeds, put $X := 1$ and stop; otherwise, with probability $1 - p_1$, simulate X^* with distribution $\left(\frac{p_2}{1-p_1}, \ldots, \frac{p_m}{1-p_1}\right)^\top$. The expected number of bits needed to specify X is the expected number for the first trial plus the expected number for the rest of the task.

Next, we quantify the notion that $h(X)$ should be small if X is nearly degenerate.

Theorem 2.9.3: *Fano's inequality. Suppose* $\# \mathcal{X} = m$ *and that the distribution of* X *is such that one value has probability* $1 - \varepsilon$. *Then*

$$h(X) \leq g(\varepsilon) + \varepsilon \log(m-1)$$
$$\leq 1 + \varepsilon \log(m-1), \tag{4}$$

where

$$g(\varepsilon) = -\varepsilon \log_+ \varepsilon - (1-\varepsilon) \log_+ (1-\varepsilon). \tag{5}$$

Proof. Set $p_1 := 1 - \varepsilon$ in (3) and bound the final h-function by $\log(m-1)$.$\square$

Joint entropy

Jointly distributed discrete random variables X, Y, taking values in $\mathcal{X}$, $\mathcal{Y}$ respectively, can be considered as a single r.v. $(X,Y)^\top$ taking values in $\mathcal{X} \times \mathcal{Y}$ with distribution $p(x,y) := P(X = x, Y = y)$, say. Its entropy

$$h(X,Y) := -\sum_x \sum_y p(x,y) \log_+ p(x,y) = -E \log_+ p(X,Y)$$

is the *joint entropy* of X and Y.

Replace X in Theorem 2.9.2 by $(X,Y)^\top$ and take ϕ to be the projection $(X,Y)^\top \mapsto X$. You get an inequality whose condition for equality is that $P((X,Y)^\top = f(X)) = 1$ for some f, or equivalently $P(Y = g(X)) = 1$ for some g. It is natural to refer to this as 'Y is with probability 1 a function of X'. You conclude

$$h(X,Y) \geq h(X), \text{ with equality iff } Y \text{ is w.p. 1 a function of } X. \tag{6}$$

Exercises

1. Suppose X takes positive-integer values, $P(X = n) = p_n > 0$ for $n = 1, 2, \ldots$, and that the p_n are non-increasing: $p_1 \geq p_2 \geq \cdots$.
    (a) Show that $h(X) < \infty$ implies $E \log X < \infty$.
    (b) By considering p_n of the form $c/(n \log^2 n)$, or otherwise, show that $E \log X < \infty$ does not imply $h(X) < \infty$.

2. Extending the pooling inequalities, consider the effect of making a probability distribution $(p_1, \ldots, p_m)^\top$ more uniform in the following sense. Suppose $p_1 > p_2$ and let $p_1' = p_1 - \varepsilon$, $p_2' = p_2 + \varepsilon$ where ε is positive and so small that $p_1' \geq p_2'$. Let $p_i' = p_i$ for $i > 2$. Show that $h(p_1', \ldots, p_m') \geq h(p_1, \ldots, p_m)$.

3. Generalise (3) by proving that, for $0 \leq m \leq n$,

$$h(p_1, \ldots, p_n) = h(p_1, \ldots, p_m, q_m) + q_m h\left(\frac{p_{m+1}}{q_m}, \ldots, \frac{p_n}{q_m}\right),$$

where $q_m = p_{m+1} + \cdots + p_n$. Deduce that

$$h(p_1, \ldots, p_n) \leq h(p_1, \ldots, p_m, q_m) + q_m \log(n - m).$$

When does equality hold?

4.   Let X be a r.v. having m possible values, with probabilities $p_1 \leq p_2 \leq \cdots \leq p_m$. Show that
(a) $h(X) \geq \sum_1^m p_k(1 - p_k) \log e \geq (1 - p_m) \log e$;
(b) $h(X) \geq -p_m \log p_m - (1 - p_m) \log(1 - p_m)$;
(c) $h(X) \geq -\log p_m$;
(d) $h(X) \geq 2(1 - p_m) \log e$.
Hint for (d): use (b) when $p_m \geq \frac{1}{2}$ and (c) when $p_m \leq \frac{1}{2}$.
     Which of the bounds in (a) and (d) is stronger and why?

5.   A real-valued function f defined on some interval I in the real line is called *convex* if

$$f(\lambda x + (1 - \lambda)y) \leq \lambda f(x) + (1 - \lambda)f(y) \qquad (x, y \in I; \, 0 \leq \lambda \leq 1),$$

and *concave* if f is convex.
(a) Suppose f is continuous on I and has non-decreasing derivative on the interior of I. By applying the Mean-value Theorem to f on the intervals joining $\lambda x + (1 - \lambda)y$ to each of x, y, show that f is convex.
(b) Use (a) to verify that g defined in (5) is concave.
(c) Prove by induction that if $x_1, \ldots, x_n \in I$ and $(p_1, \ldots, p_n)^\top$ is a probability distribution then

$$f\left(\sum_1^n p_j x_j\right) \leq \sum_1^n p_j f(x_j)$$

(a version of *Jensen's inequality*). Put this in random-variable terms.

6.   Show that $f(t) := t \log t$ is a convex function (defined in Exercise 5). Deduce that $h(p) = -\sum_1^m p_j \log p_j$ is a *concave* function of probability distributions on $\{1, \ldots, m\}$:

$$h(\lambda p + (1 - \lambda)p') \geq \lambda h(p) + (1 - \lambda)h(p') \qquad (0 \leq \lambda \leq 1). \qquad (7)$$

Here $p = (p_1, \ldots, p_m)^\top$, $p' = (p'_1, \ldots, p'_m)^\top$, and by $\lambda p + (1 - \lambda)p'$ is meant $(r_1, \ldots, r_m)^\top$ where $r_i = \lambda p_i + (1 - \lambda)p'_i$.

7.   *Continuation.* Prove (7) by the following alternative method: apply the Gibbs inequality (Lemma 1.4.1) to p and $\lambda p + (1 - \lambda)p'$, and to p' and $\lambda p + (1 - \lambda)p'$. Show that this gives a bonus: if p is not

the same as p', and $0 < \lambda < 1$, then there is strict inequality in (7). This is *strict concavity* of h.

8.   Let f be an arbitrary real-valued function on $[1, \infty)$. Define the *f-entropy* of a r.v. X with m possible values and with distribution $(p_1, \ldots, p_m)^\top$ by

$$h_f(X) := \sum_{k=1}^{m} p_k f\left(\frac{1}{p_k}\right).$$

If f is concave (see Exercise 5) show that $h_f(X) \le f(m)$.

9.   Find $(p_1, p_2, p_3)^\top$ and $(q_1, q_2, q_3)^\top$, two distinct probability distributions on an alphabet of 3 letters, such that $\sum_1^3 p_i \log p_i = \sum_1^3 q_i \log q_i$.

2.10 Conditional entropy

Let X and Y be discrete r.v.s arising from a common experiment ('jointly distributed'). If we fix some x for which $P(X = x) > 0$ we can define the conditional p.m.f. of Y given $X = x$ as $f(\cdot|x)$ given by

$$f(y|x) := P(Y = y|X = x) \qquad (y \in \mathcal{Y}).$$

It is a p.m.f. on $\mathcal{Y}$ so has an expectation

$$E(Y|X = x) := \sum_y y f(y|x),$$

the *conditional expectation of Y given $X = x$*. (If $\mathcal{Y}$ is countably infinite one must impose $E|Y| < \infty$ as a condition.) The most useful identity for conditional expectation is

$$\sum_x E(Y|X = x)P(X = x) = E(Y), \tag{1}$$

with an obvious proof.

What this is saying is that $g(x) := E(Y|X = x)$ is a real-valued function of x with the property that $E(g(X)) = E(Y)$. The memorable way to write this

$$E(E(Y|X)) = E(Y),$$

and in advanced treatments of probability it is in fact the latter which is the fundamental identity, its components carefully defined.

The *conditional entropy* of Y given X is

$$\begin{aligned} h(Y|X) &:= -E \log_+ f(Y|X) \\ &= -\sum_x \sum_y f(x, y) \log_+ f(y|x) \tag{2} \\ &= \sum_x p_X(x)\left(-\sum_y f(y|x) \log_+ f(y|x)\right). \tag{3} \end{aligned}$$

Think of this as the expected number of bits required to specify Y if X is known. The last right-hand side also expresses it as the entropy of Y given a particular value of X, averaged over the range of possible values of X.

$h(Y|X)$ is also called the *equivocation*, of X about Y.

The value of H, the information rate, that we calculated for the simple Markov source, can now be identified as $h(U_{t+1}|U_t)$ (the same for all t).

Proposition 2.10.1.
$$h(X) + h(Y|X) = h(X,Y) = h(Y) + h(X|Y).$$

Proof. The first identity comes from $\log_+ p_X(x) + \log_+ f(y|x) = \log_+ p(x,y)$; the second follows by symmetry. □

Theorem 2.10.2.
$$h(Y|X) \geq 0, \text{ with equality iff } Y \text{ is w.p. 1 a function of } X; \quad (4)$$
$$h(Y|X) \leq h(Y), \text{ with equality iff } X, Y \text{ independent}. \quad (5)$$

Proof. (4) follows from (2.9.6). For (5), take any x with $p_X(x) > 0$, then by the Gibbs inequality (Lemma 1.4.1)
$$-\sum_y f(y|x) \log_+ f(y|x) \leq -\sum_y f(y|x) \log_+ p_Y(y), \quad (6)$$
with equality iff $f(\cdot|x) = p_Y(\cdot)$ identically. Multiply by $p_X(x)$ and add, and you obtain (5) on employing (3). There is equality iff you have equality in each case of (6), i.e. $f(\cdot|x) = p_Y(\cdot)$ for every x with $p_X(x) > 0$, and that is independence. □

Corollary 2.10.3.
$$h(X,Y) \leq h(X) + h(Y), \text{ with equality iff } X \text{ and } Y \text{ are independent}.$$

Conditional independence
One says that discrete r.v.s X and Z are *conditionally independent given* Y if
$$P(X = x, Z = z|Y = y) = P(X = x|Y = y)P(Z = z|Y = y)$$
for all x, z and all y such that $P(Y = y) > 0$. A useful fact is that this is equivalent to *redundant conditioning*: the conditional p.m.f. of Z given Y and X is the same as that given just Y, or
$$P(Z = z|Y = y, X = x) = P(Z = z|Y = y).$$
Another way of thinking of this is that X, Y and Z form (3 steps of) a *Markov chain*.

Similarly to (5) one may prove (Exercise 2) that

$$h(Z|Y) \geq h(Z|Y, X), \tag{7}$$

with equality iff X and Z are conditionally independent given Y, and

$$h(Y|X) \leq h(Y|\psi(X)), \tag{8}$$

with equality iff X and Y are conditionally independent given $\psi(X)$.

Note that if we set $X^{(n)} := (X_1, \ldots, X_n)^\top$, $Y^{(n)} := (Y_1, \ldots, Y_n)^\top$ then

$$h(X^{(n)}) = \sum_1^n h(X_t|X^{(t-1)}) \leq \sum_1^n h(X_t). \tag{9}$$

Here $X^{(0)}$ is a 'blank', so $h(X_1|X^{(0)})$ is just $h(X_1)$. The equality in (9) is the *chain rule* for entropy and expresses $h(X^{(n)})$ very naturally as the sum of incremental numbers of bits needed to specify each X_t if $X^{(t-1)}$ is known.

We also have

$$h(Y^{(n)}|X^{(n)}) \leq \sum_1^n h(Y_t|X^{(n)}) \leq \sum_1^n h(Y_t|X_t), \tag{10}$$

where for the left inequality you just re-work (9) conditionally on $X^{(n)}$, while for the right-hand inequality you use (8) on each summand.

Theorem 2.10.4: the extended Fano inequality. *If X takes values $x_1, \ldots, x_m$ and Y takes values $y_1, \ldots, y_m$ and*

$$\sum_1^m P(X = x_j, Y = y_j) = 1 - \varepsilon$$

(i.e. the probability that X and Y take corresponding values is $1 - \varepsilon$), then

$$h(Y|X) \leq g(\varepsilon) + \varepsilon \log(m - 1)$$
$$\leq 1 + \varepsilon \log(m - 1).$$

Proof. Put $p_j := P(X = x_j)$, $\varepsilon_j := 1 - P(Y = y_j|X = x_j)$, so that $\sum p_j \varepsilon_j = \varepsilon$. Then, by (3) and (4),

$$h(Y|X) \leq \sum p_j(g(\varepsilon_j) + \varepsilon_j \log(m - 1))$$
$$= \sum p_j g(\varepsilon_j) + \varepsilon \log(m - 1).$$

Now g, which was defined in (2.9.5) and is sketched in Fig. 2.10.1, is concave so $\sum p_j g(\varepsilon_j) \leq g(\sum p_j \varepsilon_j) = g(\varepsilon)$ as required. (If convexity and concavity are not familiar to you, doing Exercise 5 will make sense of the last step.) ☐

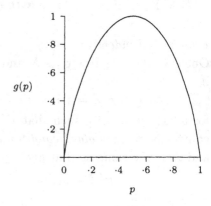

Figure 2.10.1

It is worth noting the following heuristic for the last result: the equivocation $h(Y|X)$ of X about Y is bounded by the sum of two terms. The first term $g(\varepsilon)$ is the information in the decision whether X and Y correspond or not. Assuming they do not, the second term $\varepsilon \log(m-1)$ is the maximum possible information in deciding the value of Y from the $m-1$ non-corresponding values.

Exercises

1. Let X and Y be r.v.s each taking finitely many values, and set $Z := X + Y$.
   (a) Show that $h(Z|X) = h(Y|X)$. Deduce that if X and Y are independent then $h(Z) \geq \min(h(X), h(Y))$. Give an example where the inequality is strict.
   (b) Give an example of dependent X, Y such that $h(Z) < \min(h(X), h(Y))$.

2. Prove (7) and (8).

3. Prove the following extension of Prop. 2.10.1:
$$h(X, Z|Y) = h(X|Y) + h(Z|X, Y).$$
   Deduce that
$$h(X, Z|Y) \leq h(X|Y) + h(Z|Y)$$
   with equality iff X and Z are conditionally independent given Y, thus extending Corollary 2.10.3.

2.11 The uncertainty of a stationary source

We return to considering a source that emits, or if you like *is*, a sequence $U_1, U_2, \ldots$ of r.v.s taking values in the finite alphabet $F_a = \{1, \ldots, a\}$.

Definition. *The source is* stationary *if for every* n, $k \in \mathbb{N}$, $u_1, \ldots, u_n \in F_a$,
$$P(U_1 = u_1, \ldots, U_n = u_n) = P(U_{k+1} = u_{k+1}, \ldots, U_{k+n} = u_{k+n}).$$

So the whole probability structure is unaffected by a time-shift. Note that stationarity implies
$$P(U_n = u|U_{n-1} = u_1, \ldots, U_{n-j} = u_j)$$
$$= P(U_{n+k} = u|U_{n+k-1} = u_1, \ldots, U_{n+k-j} = u_j)$$
for all $n > j$ and all k, u, u_1, $\ldots$, u_j. So stationarity of a Markov chain implies stationary transition probabilities. Conversely, a finite Markov chain

(U_n) with stationary transition probabilities is itself stationary provided you use a stationary distribution as its initial distribution (Exercise 2).

A handy notation is to write

$$\mathbf{1}\{\ \} := \begin{cases} 1, & \text{if statement } \{\ \} \text{ is true,} \\ 0, & \text{if not.} \end{cases}$$

Then for instance $\mathbf{1}\{U_1 = u_1\}$ is an *indicator* r.v. that indicates the occurrence of the event $U_1 = u_1$ by taking value 1. We shall allow ourselves to write $\mathbf{1}_{U_1=u_1}$, and similarly for other arguments, where space permits.

Our definition of stationarity can be put in terms of indicators as

$$E(\mathbf{1}\{U_1 = u_1,\ldots,U_n = u_n\}) = E(\mathbf{1}\{U_{k+1} = u_{k+1},\ldots,U_{k+n} = u_{k+n}\}),$$

which suggests that a similar identity may follow for expectations of more general r.v.s. That is indeed the case, and we will discuss it in the subsection 'The Ergodic Theorem' below, which is, however, omittable on a first reading.

To make use of stationarity we recall our notation $U^{(n)} = (U_1,\ldots,U_n)^\top$ and define

$$J_n := n^{-1}h(U^{(n)}),$$

$$K_n := \begin{cases} h(U_{n+1}|U^{(n)}) & (n = 1,2,\ldots) \\ h(U_1) & (n = 0) \end{cases}.$$

By Proposition 2.10.1,

$$K_n = h(U^{(n+1)}) - h(U^{(n)}) \qquad (n = 1,2,\ldots). \tag{1}$$

So J_n is per-letter average source entropy, while K_n is a per-letter *increment* of entropy.

Definition. If $\lim_{n\to\infty} K_n$ exists it is called the uncertainty of the source.

Theorem 2.11.1. *For a stationary source with a finite alphabet, J_n and K_n are non-increasing and have a common limit J_∞, the source uncertainty.*

Proof.

$$\begin{aligned} K_n &= h(U_{n+1}|U_1, U_2,\ldots,U_n) \\ &\le h(U_{n+1}|U_2,\ldots,U_n) && \text{by (2.10.7)} \\ &= h(U_n|U_1,\ldots,U_n) && \text{by stationarity} \\ &= K_{n-1}. \end{aligned}$$

So $K_n \downarrow$ and, being non-negative, has limit K, say. Then

$$J_n = (K_0 + K_1 + \cdots + K_{n-1})/n \qquad \text{(by (1))}$$

$$\to K \qquad \text{(Cesàro average: see Exercise 2.7.2).}$$

Also

$$J_{n+1} - J_n = \frac{nK_n - (K_0 + K_1 + \cdots + K_{n-1})}{n(n+1)} \le 0. \qquad \square$$

As we noted in §1.5, the fact that J_n decreases is what makes coding the source output by (long) segments or blocks worthwhile.

The Ergodic Theorem

We quote one more major theorem of probability theory, of relevance here and below. We need some of the terminology of ergodic theory.

Recall that in §2.1 we gave the canonical set-up for the infinite sequence of r.v.s U_1, U_2, ..., involving a probability space $(\Omega, \mathcal{A}, P)$ where $\Omega = F_a^{\mathbb{N}}$ is the space of all sequences $\omega = (\omega_1, \omega_2, \ldots)$ with each $\omega_n \in F_a$. The U_n are then coordinate-projection functions, $U_n(\omega) := \omega_n$. A general random variable on this space is any function $\phi : \Omega \to \mathbb{R}$ that has the comforting property that every set $\{\omega : \phi(\omega) \le x\}$, for $x \in \mathbb{R}$, is an event, i.e. a member of the class $\mathcal{A}$. We can thus speak of 'the event that $\phi \le x$', or, by intersection and complementation, 'the event that $x < \phi \le y$' and so on.

The *shift operator* $T : \Omega \to \Omega$ is defined by

$$T(\omega_1, \omega_2, \ldots) := (\omega_2, \omega_3, \ldots).$$

So $T^k \omega = (\omega_{k+1}, \omega_{k+2}, \ldots)$. If ϕ is any r.v. then $\phi \circ T^k$ is the composition of a k-fold shift followed by ϕ:

$$\phi \circ T^k(\omega) := \phi(T^k \omega).$$

If for instance ϕ is $\mathbf{1}\{U_1 = u\}$ then $\phi \circ T^k$ is $\mathbf{1}\{U_{k+1} = u\}$.

Stationarity formalizes the idea that the source is the same probabilistically after any time-shift, and the maximal way to express this is that for any bounded r.v. ϕ, and any $k \in \mathbb{N}$,

$$E(\phi \circ T^k) = E(\phi). \qquad (2)$$

This is in fact the standard definition of stationarity, not being restricted to discrete cases. For our discrete case it is equivalent to our definition earlier, but obviously brings the many consequences of stationarity much more to the fore, as ϕ could depend on infinitely many of its arguments. Taking ϕ to be $\mathbf{1}\{U_1 = u_1, \ldots, U_n = u_n\}$ reduces (2) to the earlier definition.

Now the Ergodic Theorem, which includes the SLLN, applies to a general stationary set-up, without our discreteness assumptions (and therefore under suitably extended definitions of random variable etc.). It says that under stationarity the time-average of the results of successive shifts settles down with probability 1 to a limit.

Theorem 2.11.2: the (Pointwise) Ergodic Theorem (G. D. Birkhoff). *Under* *stationarity, for any r.v. ϕ such that $E|\phi| < \infty$,*

$$\frac{1}{n}\sum_{k=0}^{n-1}\phi\circ T^k \xrightarrow{\text{a.s.}} \hat{\phi} \qquad (n\to\infty),$$

where $\hat{\phi}$ is a r.v. with $E|\hat{\phi}| < \infty$, $E(\hat{\phi}) = E(\phi)$, and which is invariant under T in the sense that $P(\hat{\phi}\circ T = \hat{\phi}) = 1$.

For proof see e.g. Billingsley (1965). Under the same conditions, L^1 convergence also holds: the L^1 *Ergodic Theorem*.

Exercises

1. Consider a stationary information source (U_t) with an alphabet of two symbols, a and b. Show that $P(U_{t-1} = a, U_t = b) = P(U_{t-1} = b, U_t = a)$ for all t.

2. Prove that a finite Markov chain with stationary transition probabilities is *stationary* provided you use a stationary distribution as its initial distribution.

3. Show that $n^{-1}h(U_{n+1}, \ldots, U_{2n}|U^{(n)})$ is non-increasing in n and converges to J_∞.

2.12 Ergodic sources

Definition. *A source emitting U_1, U_2, $\ldots$, with discrete alphabet F_a, is ergodic if it is stationary and, for every $j \in \mathbb{N}$, $u_1, \ldots, u_j \in F_a$,*

$$\frac{1}{n}\sum_{k=0}^{n-1}\mathbf{1}\{U_{k+1} = u_1, \ldots, U_{k+j} = u_j\}$$
$$\xrightarrow{\text{a.s.}} P(U_1 = u_1, \ldots, U_j = u_j) \qquad (n\to\infty). \tag{1}$$

Ergodicity says that, for every pattern of output the source can produce, it *will* do so, with asymptotically the right frequency. The effect of the initial conditions dies out, in a strong sense.

Again, the above is a narrow definition. A maximal characterisation, equivalent to the above, is that the stationary source must be such that every r.v. that is invariant in the sense mentioned in the Ergodic Theorem is degenerate. Alternatively, the $\hat{\phi}$ in the Ergodic Theorem has to be always degenerate at value $E(\phi)$, so, for any r.v. ϕ with $E|\phi| < \infty$,

$$\frac{1}{n}\sum_{k=0}^{n-1}\phi\circ T^k \xrightarrow{\text{a.s.}} E(\phi) \qquad (n\to\infty). \tag{2}$$

There are two sufficient conditions for ergodicity of a stationary source which seem at first sight to be weaker properties. The first is (1) with only convergence in probability:

$$\frac{1}{n} \sum_{k=0}^{n-1} \mathbf{1}\{U_{k+1} = u_1, \ldots, U_{k+j} = u_j\}$$

$$\overset{P}{\to} P(U_1 = u_1, \ldots, U_j = u_j) \qquad (n \to \infty). \qquad (3)$$

The reason this is enough is that by the Ergodic Theorem the left-hand side converges a.s. to some limit r.v.; then the convergence in probability identifies the limit so (1) follows.

The second sufficient condition in place of (1) is that for every $i, j \in \mathbb{N}$, $u_1, \ldots, u_j, v_1, \ldots, v_i \in F_a$,

$$\frac{1}{n} \sum_{k=0}^{n-1} P(U_1 = v_1, \ldots, U_i = v_i, U_{k+1} = u_1, \ldots, U_{k+j} = u_j)$$

$$\to P(U_1 = v_1, \ldots, U_i = v_i) P(U_1 = u_1, \ldots, U_j = u_j) \qquad (n \to \infty). (4)$$

For proof that this suffices see Billingsley (1965), Theorem 1.4. Note that it is *not* enough that

$$\frac{1}{n} \sum_{k=0}^{n-1} P(U_{k+1} = u_1, \ldots, U_{k+j} = u_j) \to P(U_1 = u_1, \ldots, U_j = u_j).$$

The Shannon-McMillan-Breiman Theorem

This is the main result about ergodic sources. Recall $p_n(u) := P(U^{(n)} = u)$ and $\xi_n := -\frac{1}{n} \log_+ p_n(U^{(n)})$.

Theorem 2.12.1: Shannon-McMillan-Breiman Theorem. *For an ergodic source, ξ_n converges a.s. and in L^1 to J_∞, the source uncertainty.*

The L^1 convergence was established by C. E. Shannon and B. McMillan in 1948–53, and the extension to a.s. convergence by L. Breiman in papers of 1957 and 1960. The best proof of the full result is that by Algoet & Cover (1988). It uses probabilistic machinery more advanced than that which we have developed here, so we give a modification of the elementary proof in Thomasian (1960), of the L^1 convergence only. Our proof hinges on one lemma, a sort of converse of the First Coding Theorem. Recall the idea of 'reliable encoding rate' from §2.2.

Lemma 2.12.2. *Suppose the source is reliably encodable at every rate $r > J_\infty$. Then $\xi_n \overset{L^1}{\to} J_\infty$.*

Proof. Choose $r > r' > J_\infty$. The assumption gives sets $A_n \subseteq F_a^n$ with $\# A_n \le 2^{nr'}$ and $P(U^{(n)} \in A_n) \to 1$. Now

$$P(U^{(n)} \in A_n) \le P(\xi_n \le r) + P(U^{(n)} \in A_n, \xi_n > r).$$

The latter probability is the sum of at most $2^{nr'}$ probabilities $P(U^{(n)} = u) = p_n(u)$, each of which has $-\frac{1}{n}\log p_n(u) > r$, or equivalently $p_n(u) < 2^{-nr}$. Thus

$$P(U^{(n)} \in A_n) \leq P(\xi_n \leq r) + 2^{nr'}2^{-nr}.$$

Since the left-hand side tends to 1 and $r > r'$ it follows that $P(\xi_n \leq r) \to 1$.

We thus have $q_n := P(\xi_n > r) \to 0$ and need next to strengthen this to $E(\xi_n \mathbf{1}\{\xi_n > r\}) \to 0$. Let $B := \{u \in F_a^n : -\frac{1}{n}\log p_n(u) > r\}$, and observe that $(p_n(u)/q_n)_{u \in B}$ is a probability distribution on B, hence, by Proposition 2.9.1,

$$-\sum_{u \in B} \frac{p_n(u)}{q_n} \log \frac{p_n(u)}{q_n} \leq \log \# B.$$

The right-hand side is at most $\log \# F_a^n = n \log a$. Inserting this and rearranging, you find that

$$\sum_{u \in B} (-n^{-1}\log p_n(u))p_n(u) \leq q_n \log a - n^{-1}q_n \log q_n.$$

The right-hand side tends to 0 and the left-hand side is $E(\xi_n \mathbf{1}\{\xi_n > r\})$ which, being non-negative, does thus tend to 0 as claimed.

Next,

$$E|\xi_n - J_\infty| + E(\xi_n - J_\infty)$$
$$= 2E((\xi_n - J_\infty)\mathbf{1}\{\xi_n > J_\infty\})$$
$$= 2E((\xi_n - J_\infty)\mathbf{1}\{J_\infty < \xi_n \leq r\}) + 2E((\xi_n - J_\infty)\mathbf{1}\{\xi_n > r\}).$$

On the right the first term is at most $2(r - J_\infty)$, while the second term tends to 0 because it is $2E(\xi_n \mathbf{1}\{\xi_n > r\}) - 2J_\infty P(\xi_n > r)$. Since $E(\xi_n) = \frac{1}{n}h(U^{(n)}) = J_n \to J_\infty$ you conclude that

$$\limsup_{n \to \infty} E|\xi_n - J_\infty| \leq 2(r - J_\infty).$$

Since r can be set to any value exceeding J_∞ this implies $E|\xi_n - J_\infty| \to 0$, as required. $\square$

Proof of Theorem 2.12.1. (L^1 convergence only.) To satisfy Lemma 2.12.2 we pick $r > J_\infty$ and will then need to define various quantities in relation to r, namely an integer $m \geq 2$, a constant $0 < \delta < 1$ and an integer $n_0 \geq m$. For the present take these as given. We shall write elements of F_a^k as strings $s = s_1, \ldots, s_k$ for various k.

For any $n \geq m$, any $s \in F_a^m$ and any $u = u_1, \ldots, u_n \in F_a^n$, let $N_s(u)$ be the number of times the substring s appears in the string u, so

$$N_s(u) := \sum_{k=0}^{n-m} \mathbf{1}\{u_{k+1}, \ldots, u_{k+m} = s\}.$$

Let G be the set of 'good' m-strings $s \in F_a^m$, those with $P(U^{(m)} = s) > 0$. Let B_n be the set of all $u \in F_a^n$ such that, for every 'bad' string $s \in F_a^m \setminus G$, you have $N_s(u) = 0$. By stationarity, for a 'bad' s we have $P(U_{k+1}, \ldots, U_{k+m} = s) = 0$ for every k. So $P(U^{(n)} \in B_n) = 1$.

Let C_n be the set of all $u \in F_a^n$ such that, for all $s \in G$,

$$n^{-1}N_s(u) < P(U^{(m)} = s) + \delta.$$

By ergodicity, for each s,

$$(n - m)^{-1} N_s(U^{(n)}) \xrightarrow{\text{a.s.}} P(U^{(m)} = s) \qquad (n \to \infty),$$

and the same therefore holds when the divisor $n - m$ is replaced by n. So, with probability 1, for each $s \in G$ you have $\frac{1}{n} N_s(U^{(n)}) < P(U^{(m)} = s) + \delta$ for all large n. Thus $P(U^{(n)} \in C_n) = 1$ for all large n. So with $A_n := B_n \cap C_n$ you have $P(U^{(n)} \in A_n) = 1$ for all large n. This sequence of sets A_n is our choice for the reliable-encoding-at-rate-r definition, so we prove $\# A_n \le 2^{nr}$ for all large n.

Let Q be another probability distribution for $U_1, U_2, \ldots$, defined by

$$Q(U_1 = u_1, \ldots, U_{m-1} = u_{m-1}) = P(U_1 = u_1, \ldots, U_{m-1} = u_{m-1}),$$

and, for $n \ge m$,

$$Q(U_n = u_n | U_1 = u_1, \ldots, U_{n-1} = u_{n-1})$$
$$= \begin{cases} P(U_n = u_n | U_{n-m+1} = u_{n-m+1}, \ldots, U_{n-1} = u_{n-1}), & \text{if } u_1, \ldots, u_n \in B_n, \\ 0, & \text{otherwise.} \end{cases}$$

Note that the right-hand side is well-defined. If you multiply through by $Q(U_1 = u_1, \ldots, U_{n-m} = u_{n-m})$ and add over all $u_1, \ldots, u_{n-m}$ the right-hand side is unaltered so you obtain

$$Q(U_n = u_n | U_{n-m+1} = u_{n-m+1}, \ldots, U_{n-1} = u_{n-1})$$
$$= P(U_n = u_n | U_{n-m+1} = u_{n-m+1}, \ldots, U_{n-1} = u_{n-1}).$$

Thus in defining Q we are taking the mth-order transition probabilities of the source and using them to define an $(m-1)^{\text{th}}$-order Markov chain.

For any $s \in F_a^{m-1}$ and $j \in F_a$ we write $sj \in F_a^m$ for their concatenation. Set

$$q_s := P(U^{(m-1)} = s) = Q(U^{(m-1)} = s) \qquad (s \in F_a^{m-1}),$$
$$q_{j|s} := Q(U_m = j | U^{(m-1)} = s) = P(U_m = j | U^{(m-1)} = s) \qquad (sj \in G).$$

Then for any $u = u_1, \ldots, u_n \in B_n$, writing u' for the initial substring $u_1, \ldots, u_{m-1}$,

$$Q(U^{(n)} = u) = q_{u'} \prod_{sj \in G} (q_{j|s})^{N_{sj}(u)}.$$

For any $sj \in G$ you have $P(U^{(m)} = sj) = q_s q_{j|s}$, so the inequality defining C_n may be written $N_{sj}(u) < n(q_s q_{j|s} + \delta)$. Putting this bound into the above you conclude that, for every $u = u_1, \ldots, u_n \in A_n$,

$$Q(U^{(n)} = u) \ge q_{u'} \prod_{sj \in G} (q_{j|s})^{n(q_s q_{j|s} + \delta)}.$$

Now $-\sum_{sj \in G} q_s q_{j|s} \log q_{j|s} = h(U_m | U^{(m-1)}) = K_{m-1}$. So the bound may be written

$$Q(U^{(n)} = u) \ge \left(q_{u'}^{1/n} \prod_{sj \in G} q_{j|s}^{\delta} \right)^n 2^{-K_{m-1}n}.$$

It now becomes clear what the recipe for choosing the constants should be. Since $K_{m-1} \to J_\infty$ we can fix m so large that $K_{m-1} < r$. Set $\varepsilon := r - K_{m-1} > 0$. Choose $\delta > 0$ so small that $\left(\prod_{sj \in G} q_{j|s} \right)^{\delta} \ge 2^{-\varepsilon/2}$. Finally choose $n_0 > m$ so large that for each of the finite set of $(m-1)$-strings $u_1, \ldots, u_{m-1}$ with $q_{u_1, \ldots, u_{m-1}} > 0$ you have

$q_{u_1,\ldots,u_{m-1}}^{1/n_0} \geq 2^{-\varepsilon/2}$. Then, for $n \geq n_0$,

$$Q(U^{(n)} = u) \geq \left(2^{-\varepsilon/2}2^{-\varepsilon/2}\right)^n 2^{-K_{m-1}n} = 2^{-nr} \qquad (u \in A_n),$$

so

$$1 \geq \sum_{u \in A_n} Q(U^{(n)} = u) \geq 2^{-nr} \, \# \, A_n,$$

and $\# \, A_n \leq 2^{nr}$ as required.   □

The L^1 convergence that we have just proved is enough for our purposes. In particular it implies convergence in probability. So, by the FCT,

$$\boxed{H = J_\infty}$$

and we can *identify the uncertainty and the information rate*. Also *every ergodic source satisfies the AEP*.

For Bernoulli and 'good' Markov sources we proved $\xi_n \xrightarrow{\text{P}} J_\infty$ directly, in §§2.5, 2.7. Bernoulli sources are obviously ergodic. A simple Markov source needs its initial distribution to be a stationary distribution, for stationarity, and then if irreducible is ergodic (Exercise 2). So, apart from the minor extension of allowing the initial distribution of the Markov chain to be non-stationary, our results for Bernoulli and Markov sources are special cases of the Shannon-McMillan-Breiman Theorem.

Failure of ergodicity

Ergodicity is long-term lack of dependence. Its easiest failure is for a source defined as follows. Toss a coin. If you get Heads, let the source emit 1s, forever; if Tails, let the source emit 0s forever. You might as well call the result of the toss ω. Then

$$\frac{1}{n}\sum_1^n \mathbf{1}\{U_k = 1\} = \begin{cases} 1 & \text{if } \omega = H \\ 0 & \text{if } \omega = T \end{cases}.$$

The right-hand side is a r.v., taking values 0 and 1 with probabilities $\frac{1}{2}$. So the left-hand side is fixed at one same r.v., and does not converge in any sense to anything non-random.

For an example with more content consider the following.

Example 2.12.3. Suppose that with probability α_j the source letters are independent and equidistributed in an alphabet of a_j letters. Here the α_j are positive and $\sum_1^k \alpha_j = 1$. The alphabets $\mathcal{A}_1, \ldots, \mathcal{A}_k$ are distinct, and $k > 1$. Then the left-hand side of (2.12.2) evaluates, if we set $\phi := \mathbf{1}\{U_1 \in \mathcal{A}_j\}$, as

$$\frac{1}{n}\sum_1^n \phi \circ T^k = \frac{1}{n}\sum_{k=1}^n \mathbf{1}\{U_{k+1} \in \mathcal{A}_j\} = \begin{cases} 1, & \text{with probability } \alpha_j, \\ 0, & \text{with probability } 1 - \alpha_j. \end{cases}$$

Again, this is fixed at a proper r.v., so the source is not ergodic. Let us do three calculations.

(i) $\xi_n = X$ for each n, where X is a r.v. that is $\log a_j$ with probability α_j. So ξ_n converges (in all senses!) as $n \to \infty$, but to a non-degenerate r.v. The source therefore does not satisfy the condition of the FCT, or equivalently *it does not have the AEP.*

(ii)   Since the number of n-strings from $\mathcal{A}_j$ is a_j^n, and each has probability $\alpha_j a_j^{-n}$,

$$J_n = -\frac{1}{n} \sum_j a_j^n (\alpha_j a_j^{-n}) \log(\alpha_j a_j^{-n})$$

$$= -\frac{1}{n} \sum_j \alpha_j \log \alpha_j + \sum_j \alpha_j \log a_j.$$

So $J_\infty = \sum_j \alpha_j \log a_j$.

(iii)   Because $U^{(n)}$ has to be a string from *one* of the alphabets, in evaluating $P(U^{(n)} \in A_n)$ it is enough to consider only A_n of the form $\bigcup_{j=1}^k A_{n,j}$, where $A_{n,j} \subseteq \mathcal{A}_j^n$. Then

$$P(U^{(n)} \in A_n) = \sum_{j=1}^k \alpha_j a_j^{-n} \# A_{n,j},$$

which can converge to 1 as $n \to \infty$ only if $a_j^{-n} \# A_{n,j} \to 1$ for each j. The greatest of the a_j is a_l, say. Then for r to be a reliable encoding rate you need

$$2^{nr} \geq \# A_n > \# A_{n,l} \sim a_l^n$$

as $n \to \infty$. So $nr \geq n \log a_l + o(1)$, and hence any $r > \log a_l$ will do, but no $r < \log a_l$. Finally $H = \inf r = \log a_l = \max_j \log a_j$.

The answers in (i), (ii) and (iii) are all different, while for an ergodic source they must agree.

Exercises

1.   Consider a variant of Example 2.12.3 in which $a_1 < a_2 < \cdots < a_k$ and the alphabets are nested: $\mathcal{A}_1 \subseteq \mathcal{A}_2 \subseteq \cdots \subseteq \mathcal{A}_k$, instead of being disjoint.
        (a) Show that $H = \max_j \log a_j$ as before.
        (b) To evaluate J_n, argue that there are $a_j^n - a_{j-1}^n$ n-strings using letters from $\mathcal{A}_j$ but not entirely from $\mathcal{A}_{j-1}$, and these could have come from any of $\mathcal{A}_j, \ldots, \mathcal{A}_k$ so each has probability $\alpha_j a_j^{-n} + \cdots + \alpha_k a_n^{-n}$. Deduce that $J_\infty = \sum_j \alpha_j \log a_j$ as in Example 2.12.3.

2.   Show that an irreducible simple Markov source, with its stationary distribution as initial distribution, is ergodic.

3.   Consider a simple Markov source with transition matrix $\left(\begin{smallmatrix} A & 0 \\ 0 & B \end{smallmatrix}\right)$, where
      A is an $a \times a$ irreducible aperiodic transition-matrix and B is a $b \times b$
      irreducible aperiodic transition-matrix. Show that the source is not
      irreducible. Why would you expect its uncertainty to be $(p_1 + \cdots + p_a)H_A + (p_{a+1} + \cdots + p_{a+b})H_B$, where H_A, H_B are the information
      rates of the irreducible simple Markov sources with matrices A, B
      respectively? Establish that this is indeed its uncertainty. What is
      its information rate?

4.   For an ergodic source consider the $(m-1)^{\text{th}}$-order Markov approxi-
      mation, as in the proof of the Shannon-McMillan-Breiman Theorem.
      Show that the latter is irreducible, that is, for any pair of letters
      from its alphabet, there is positive probability that having emitted
      the first letter the source will emit the second letter sometime later.

2.13 Further topics

Source coding with a fidelity criterion
Suppose that the stream U_1, U_2, ... of source symbols from alphabet F_a
is required to be expressed in a possibly different alphabet G_b, as X_1, X_2,
..., say, and that some distortion is permitted in doing so. The n-block
$U^{(n)} = (U_1, \ldots, U_n)^{\top}$ is encoded by some deterministic function g_n into an n-
block $X^{(n)} = (X_1, \ldots, X_n)^{\top} = g_n(U^{(n)})$ and the *distortion* is $d_n(U^{(n)}, X^{(n)})$
where $d_n : F_a^n \times G_b^n \rightarrow [0, \infty)$ is some given function called the *distortion
measure*. A whole suite of distortion measures, one for each n, is called a
fidelity criterion. We might for instance if $G_b \subseteq F_a$ have

$$d_n(u^{(n)}, x^{(n)}) = \mathbf{1}\{u^{(n)} \neq x^{(n)}\},$$

that is, no distortion if the input is perfectly reproduced and a fixed positive distortion if not. An important type of fidelity criterion uses *additive*
distortion-measures

$$d_n(u^{(n)}, x^{(n)}) := \sum_{i=1}^{n} d_1(u_i, x_i),$$

based upon some per-letter distortion-measure d_1. If $d_1(u, x) := \mathbf{1}\{u \neq x\}$
then the additive distortion-measures based upon it just count the number
of errors in their blocks.

   We define the *average distortion* under a chosen fidelity-criterion to be
$Ed_n(U^{(n)}, g_n(U^{(n)}))$ and the *per-letter average distortion* to be $n^{-1}Ed_n(U^{(n)}, g_n(U^{(n)}))$. If δ represents a tolerable level for this quantity then we should

restrict attention to codes g_n that do not exceed it, at least for large block-sizes, that is, for which

$$n^{-1}Ed_n(U^{(n)}, g_n(U^{(n)})) \leq \delta \quad \text{for all large } n. \tag{1}$$

We want to extend our notion 'reliably encodable at rate r' in §2.2 to some analogous 'encodable at distortion δ at rate r'. In place of the set A_n of the former definition we use $\{x : P(g_n(U^{(n)}) = x) > 0\}$, the set of essentially *all* codewords of g_n. It may well have fewer elements than the set of possible values of $U^{(n)}$, if for instance we are trying to code for a particular device with restrictions on what it can accept.

Definition. *Suppose we can find a sequence of codes g_n satisfying (1) and such that*

$$\#\{x : P(g_n(U^{(n)}) = x) > 0\} \leq 2^{n(r+o(1))} \quad (n \to \infty). \tag{2}$$

Then the source is encodable with distortion δ at rate r.

The units of r are bits per source-symbol. If $U^{(n)}$ is emitted over time n then they become bits/unit time. Analogously to the reliable-encoding case, the idea of the definition is that in principle you can encode almost at rate r with distortion at most δ for long enough blocks. 'In principle' because codes that have per-letter average distortion at most δ may be impracticable to implement.

Definition. *Let $R(\delta)$ be the infimum of all rates r at which the source is encodable with distortion δ. Then $R(\cdot)$ is the* rate-distortion function *for the source.*

If you have a device such as a fax, modem or compact-disc reading-head, that can accept input at rate $p > R(\delta)$ bits/unit time, then by suitably encoding sufficiently long blocks of source symbols you can put the source-output through the device at per-symbol average distortion no more than δ.

Straightforward properties of R are, first, that it is a non-increasing function (for if you increase δ more codes may satisfy (1), so the range of r for which a code can be found to make (2) hold will grow, or at least not lessen.) Second, if you take $g_n(u^{(n)})$ to be an $x^{(n)} \in G_b^n$ that minimizes $d_n(u^{(n)}, x^{(n)})$ then the code so defined minimizes the left-hand side of (1). Let $\delta_{\min}$ be the limsup, as $n \to \infty$, of the minimized left-hand sides. The source is then not encodable with distortion below $\delta_{\min}$. So R is well-defined on the interval $(\delta_{\min}, \infty)$, and we set $R(\delta_{\min}) := \lim_{\delta \downarrow \delta_{\min}} R(\delta)$. On the other hand a deterministic encoding with $g_n(u^{(n)})$ equal to some fixed x, for all $u^{(n)}$, gives rate 0, and if we set $\delta_{\max}$ to be $\inf_x \limsup_{n \to \infty} n^{-1}Ed_n(U^{(n)}, x)$ then it is clear that $R(\delta) = 0$ for all $\delta \geq \delta_{\max}$. The interesting range for the argument of R is thus from $\delta_{\min}$ to $\delta_{\max}$ (though $\delta_{\max}$ could be $+\infty$).

The information rate H of the source, discussed in the earlier sections of this chapter, can with a suitable fidelity criterion be identified with $R(0)$, where also $\delta_{\min} = 0$. To make the identification let the code alphabet be the same as the source alphabet and put

$$d_n(u^{(n)}, v^{(n)}) = \begin{cases} n, & \text{if } u^{(n)} \neq v^{(n)}, \\ 0, & \text{if } u^{(n)} = v^{(n)}. \end{cases}$$

Then g_n is a map from F_a^n to itself, and $n^{-1}Ed_n(U^{(n)}, g_n(U^{(n)}))$ reduces to $P(g_n(U^{(n)}) \neq U^{(n)})$, so that for distortion less than ε one needs to have $P(g_n(U^{(n)}) = U^{(n)}) \geq 1 - \varepsilon$ for all large n, as well as (2). The best g_n will have $g_n(u) = u$ for some set A_n of 2^{nr_n} distinct values u. The identification of the source being encodable with distortion less than every $\varepsilon > 0$ at rate r, and reliably encodable at rate r, is then clear: in both cases it means that we can find such sets A_n with $r_n \to r$.

For $\delta \in (\delta_{\min}, \delta_{\max})$ it is easy to obtain a lower bound on $R(\delta)$. (2) says that the distribution of $g_n(U^{(n)})$ lives on a set of at most 2^{nr_n} elements, where $r_n \to r$ as $n \to \infty$. Take n large enough for the inequality in (1) to hold. From Exercise 1.4.4 we know that $h(g_n(U^{(n)})) \leq \log 2^{nr_n} = nr_n$. Thus r_n is at least the infimum, over all g_n such that $n^{-1}Ed_n(U^{(n)}, g_n(U^{(n)})) \leq \delta$, of $n^{-1}h(g_n(U^{(n)}))$. Now $U^{(n)}$ and $X^{(n)} = g_n(U^{(n)})$ have a joint distribution in which the (marginal) distribution of $U^{(n)}$ is determined by the source while the conditional distribution of $X^{(n)}$ given $U^{(n)}$ is degenerate. If we widen the class of conditional distributions, in other words allow 'random' coding, then the said infimum cannot go up. Furthermore $h(X^{(n)})$ is at least the 'mutual information' $I(U^{(n)} \wedge X^{(n)}) = h(X^{(n)}) - h(X^{(n)}|U^{(n)})$ between $U^{(n)}$ and $X^{(n)}$, by (4.1.2). (See §4.1 for details of mutual information.) Hence if we set

$$i_n := \inf n^{-1}I(U^{(n)} \wedge X^{(n)}),$$

the infimum being over all conditional distributions for $X^{(n)}$ given $U^{(n)}$, such that $n^{-1}Ed_n(U^{(n)}, X^{(n)}) \leq \delta$, then we find that $r_n \geq i_n$. Consequently $R(\delta) \geq \limsup_{n \to \infty} i_n$.

The remarkable fact is that this lower bound is the actual *value* of $R(\delta)$. Shannon's *Source Coding Theorem* of 1959 states that for an ergodic source, subject to additive distortion-measures d_n, if $i := \lim_{n \to \infty} i_n$ exists then $R(\delta) = i$.

Many further results in source coding have followed the Shannon theorem quoted, and the subject remains in an active state of development today. Gray (1990a) is an up-to-date text devoted to it.

Epsilon-entropy, metric entropy and algorithmic information theory
All the words in the heading are associated with the name Andrei Kolmogorov, a giant of 20th century mathematics who died in 1987.

Suppose that we try to extend the definitions of our source-coding discussion to spaces larger than the finite alphabets we considered above, indeed to general metric spaces. Suppose we wish to 'code', or represent, some large subset A of a metric space S into some smaller subset B. Let m denote the metric and let ε be some positive constant. The distortion measure

$$d_\varepsilon(a,b) := \begin{cases} 0, & \text{if } m(a,b) \le \varepsilon, \\ \infty, & \text{if } m(a,b) > \varepsilon \end{cases}$$

imposes a maximum-distance constraint. That is, if X and Y are random variables taking values in A and B respectively then the only way that $Ed_\varepsilon(X,Y)$ can be finite is if every possible value of X (taken with positive probability) is within distance ε of some possible value of Y. Let us write

$$N_\varepsilon(s) := \{s' \in S : m(s,s') \le \varepsilon\} \qquad (s \in S)$$

for the closed ε-balls centred on elements of S. Let $2^{l(\varepsilon)}$ be the least integer such that there exist balls $N_\varepsilon(b_1)$, ..., $N_\varepsilon(b_{2^{l(\varepsilon)}})$, centred on elements of B, whose union contains A. Then Y must have at least $2^{l(\varepsilon)}$ distinct values, and one would need all binary $l(\varepsilon)$-strings to represent all its values. Kolmogorov's idea is that in metric spaces of interest, such as function-spaces, $l(\varepsilon)$ typically grows like $2^{r/\varepsilon}$ as $\varepsilon \downarrow 0$, for some constant r called the *exponent of metric entropy* of the set A. Its formal definition is

$$r := \limsup_{\varepsilon \downarrow 0} \frac{\log l(\varepsilon)}{\log(1/\varepsilon)}.$$

This idea has found many applications in different branches of mathematics. The exponent r represents in some sense the 'size' or 'complexity' of the set A. For instance, the class of all subsets of the d-dimensional unit cube $[0,1]^d$ is a metric space under the Hausdorff metric* d_H. Within this class the set of all polyhedra with at most m vertices has exponent of metric entropy 0, but the set of all convex figures has exponent $\frac{1}{2}(d-1)$ (Dudley (1974)). Application of 'metric entropy' to function approxima-

---

* $d_H(A,B)$ is defined as the infimum of those ε such that every point of A is within Euclidean distance ε of some point of B, and every point of B is within Euclidean distance ε of some point of A.

tion was developed in Kolmogorov & Tikhomirov (1961). An instance of
what can be proved is a theorem of Vitushkin and Kolmogorov that it is
impossible to represent a general r-times differentiable function of n real
variables in terms of q-times differentiable functions of m real variables, if
$n/r > m/q$. The survey article Lorentz (1966) gives an overview, and a
recent monograph is Carl & Stephani (1990).

We have met in this and the previous chapter the ideas that the number
of bits needed to describe the elements of a set S is $\log(\# S)$, and that the
number of bits needed to describe a random variable X is $h(X)$. Starting
with an explicit analogy with these notions, Kolmogorov defined in 1965
the 'complexity' $K(x)$ of a finite object x as the shortest binary program
that causes a 'universal computer' to print x. He was able to show that
this definition depends little on the choice of computer, and that $K(x)$ and
associated quantities $K(x,y)$, $K(y|x)$ have properties analogous to those of
$h(X)$, $h(X,Y)$, $h(Y|X)$ that we met in §§2.9–10. The name 'algorithmic
information theory' thus naturally attached itself to this notion of complexity. The theory has developed substantially in subsequent years; for an
overview the article Cover et al. (1989), one of a set of obituary articles on
Kolmogorov, can be recommended.

There also exists a theory of *computational complexity* that has developed
in parallel to the algorithmic complexity theory discussed above. Computational complexity has gained wide acceptance, in part because of its relevance to cryptography, and textbook treatments of it exist: see for example
Welsh (1988), Chapter 9.

Statistical inference
There are many close connections between information theory and statistical
inference. For a modern textbook account see Blahut (1987), Chapters 4
and 8.

Suppose $p = (p_x)$ and $q = (q_x)$ are probability distributions on some finite
set $\mathcal{X}$. The *discrimination* $D(p\|q)$ is defined by

$$D(p\|q) := \sum_{x \in \mathcal{X}} p_x \log \frac{p_x}{q_x},$$

and turns out to give a good measure of how much q is statistically distin-
guishable from p. Suppose that you have a random sample of size k (that
is, the values of k independent random variables) from a distribution that
is either p, that being the 'Null Hypothesis' H_0, or q, that being the Alternative Hypothesis H_1. A test of these hypotheses is a rule as to when to
reject H_0 in favour of H_1, based on the value of the random sample. It is

usual to insist that the probability of a 'Type I error', rejecting H_0 when it holds, is at most some pre-specified 'significance level' $\alpha > 0$. The best or 'most powerful' test is then one that, subject to this significance level, minimizes the probability of a 'Type II error', not rejecting H_0 when in fact H_1 holds. The minimized probability is

$$\beta_k(\alpha) := \min_{A \subseteq \mathcal{X},\, p^k(A) \geq 1-\varepsilon} q^k(A),$$

where $p^k(A) = \sum_{(x_1,\ldots,x_k) \in A} p_{x_1} \cdots p_{x_k}$ and similarly for q^k. (Here A is the set of values of the random sample within which the test chooses not to reject H_0.) We then have the following result:

Stein's Lemma (C. Stein, 1952). *For each $0 < \alpha < 1$ the asymptotic behaviour of the Type II error for a most powerful test is $\beta_k(\alpha) = 2^{-D(p\|q)k(1+o(1))}$ as $k \to \infty$.*

This has subsequently been extended in various ways. For instance it is possible to find a sequence of tests such that the probabilities α_k and β_k of Type I and Type II errors converge simultaneously to zero; indeed

$$\alpha_k = e^{-D(r\|p)k(1+o(1))} \quad \text{and} \quad \beta_k = e^{-D(r\|q)k(1+o(1))},$$

where r is a distribution 'between' p and q in a certain sense.

So much for hypothesis testing. There is also a connection with statistical estimation, in that Fisher's 'statistical information' $I(\theta)$, for a family of distributions $p(\theta) = (p_x(\theta))$ depending on a parameter θ, is related to discrimination by

$$\lim_{\theta' \to \theta} \frac{D(p(\theta')\|p(\theta))}{(\theta' - \theta)^2} = \frac{I(\theta)}{\ln 4}, \tag{3}$$

(see Problem 2.14.21).

On a more general note, the *Jaynes maximum-entropy principle* and Kullback's *minimum-discrimination principle* give general methods of inference having wide applicability. See Robert (1990) for a modern account.

Ergodic theory

Notions of entropy occupy a central position in the mathematical area known as *ergodic theory*, which we have briefly touched upon in our discussion of stationary sources and the Ergodic Theorem in §2.11. Billingsley (1965) gives a general account, and Ornstein & Weiss (1991) might profitably be consulted for an advanced recent exposition.

2.14 Problems

1.   Consider a Bernoulli source having probability distribution $p_a = \frac{1}{2}$, $p_b = \frac{1}{4}$, $p_c = \frac{1}{8} = p_d$ over a 4-letter alphabet $F_4 = \{a, b, c, d\}$. Take the output in blocks of length $n = 12m$, and consider a particular $u \in F_4^n$ in which the letter a occurs $5m$ times, letter b occurs $5m$ times, letter c occurs m times and letter d occurs m times. Show that u is 'weakly typical' in that $|-\frac{1}{n}\log P(U^{(n)} = u) - h| < \delta$ for all $\delta > 0$, but that $u = u_1 u_2 \ldots u_n$ does not have the property of being 'strongly typical':

$$|n^{-1} \#\{i : 1 \le i \le n, u_i = j\} - p_j| < \delta \qquad (j \in F_4).$$

2.   Consider a particular Bernoulli source that emits symbols 1, 0 with probabilities p, $1 - p$ respectively, where $0 < p < 1$. Let $h := -p \log p - (1 - p) \log(1 - p)$ and let $\varepsilon > 0$ be fixed. Let $U^{(n)}$ be the string consisting of the first n symbols emitted by the source. Prove that there is a set S_n of possible values of $U^{(n)}$ such that the probability that $U^{(n)}$ belongs to S_n is at least

$$1 - \left(\log \frac{p}{1-p}\right)^2 p(1 - p)/(n\varepsilon^2),$$

and so that for each $s \in S_n$ the probability that $U^{(n)}$ is s lies between $2^{-n(h+\varepsilon)}$ and $2^{-n(h-\varepsilon)}$. (Cambridge 1990)

3.   Write $h(p) := -\sum_1^m p_j \log p_j$ for $p = (p_1, \ldots, p_m)^\top$ a distribution on m values.
      (a) Show that $h(Ap) \ge h(p)$ if A is a doubly stochastic matrix (i.e. a square matrix of non-negative elements for which all row and column sums are unity).
          Hint: show that what you have to prove is equivalent to

$$-\sum_{i,j} a_{ij} p_j \log p_j \le -\sum_{i,j} a_{ij} p_j \log p_i',$$

where $p_i' := (Ap)_i$. Then consider the equivalent inequality obtained by adding $-\sum_{i,j} a_{ij} p_j \log a_{ij}$ to both sides.
      (b) To characterise when equality occurs in (a) observe first that $h(Ap) = h(p)$ for all p if A is a permutation matrix. Show that
          (i) if p is equidistribution then $h(Ap) = h(p)$ for all doubly-stochastic A;
          (ii) if the elements of p are not all distinct then there is a non-permutation doubly-stochastic A such that $h(Ap) = h(p)$;

(iii) if p has all its elements *distinct* and A is a doubly stochastic matrix with $h(Ap) = h(p)$, then A is a permutation matrix.

4. Consider a source with letters chosen from an alphabet of size $a + b$, for which the message strings are constrained by the condition that no two letters of A should ever occur consecutively, where A is a subset of the alphabet of size a.

(a) Suppose the message follows a Markov chain, all characters which are permitted at a given place being equally likely. Show that this source has information rate

$$H = \frac{a \log b + (a + b) \log(a + b)}{2a + b}.$$

(b) By solving a recurrence relation, or otherwise, find how many strings of length n satisfy the constraint. Suppose these are all equally likely. Show that the source has information rate

$$H = \log \left(\frac{b + \sqrt{b^2 + 4ab}}{2} \right).$$

Why are the two answers different?

5. Suppose that U_t can take values 0 or 1, and that the probability that $U_t = 1$, conditional on $U_{t-1}, U_{t-2}, \ldots$, is b_j, where j is the time that has elapsed since U last equalled 1. Assume that there is an n_0 such that $0 < b_j < 1$ for $0 < j \le n_0$ and $b_j = 1$ for $j > n_0$. Show that the process (X_t), where X_t is the time elapsed since U last equalled 1, is Markov, and calculate its information rate. (Cambridge 1977)

6. A source transmits in $1 + \sum_{j=1}^{J} m_j$ symbols, comprising J mutually exclusive alphabets, the jth having m_j letters, and a universal space character. After the space character, each letter of alphabet j has probability p_j/m_j of occurring. After any letter of alphabet j, each *other* letter of alphabet j, and the space character, has probability $1/m_j$ of occurring.

Determine the information rate of the source, and show precisely how it comprises a term attributable to the information supplied by specification of the alphabet in which a 'word' (sequence between spaces) is transmitted, and a term attributable to the information in the sequence of characters in that word. (Cambridge 1980)

7. A binary source is second-order Markov and emits digits 0 or 1

according to the rule

$$P(X_t = k | X_{t-1} = j, X_{t-2} = i) = q_r,$$

where k, j, i and r take values 0 or 1, $r = k - j - i$ (mod 2), and $q_0 + q_1 = 1$. Determine the information rate of the source.

Explain the relationship between this rate and that of a binary Bernoulli source, emitting digits 0 and 1 with probabilities q_0 and q_1. (Cambridge 1983)

8.   At each time-unit a device reads the current version of a string of k characters each of which may be either 0 or 1. It then transmits the number of characters that are equal to 1. Between each reading the string is perturbed by changing one of the characters at random (from 0 or 1 or vice versa, with each character being equally likely to be changed). Determine an expression for the information rate of this source. (Cambridge 1985)

9.   The daily discretized meteorological observations at the Downunder Antarctic Base can be assumed to constitute an irreducible aperiodic Markov chain with known transition matrix. However, the base strictly observes the weekend, and radios the day's readings through only on Monday to Friday. Determine the information rate of the transmitted readings. (Assume that the formula for the information rate of a Markov source holds also for periodic chains.) (Cambridge Dipl. Stat. 1986)

10.   A Bernoulli source having information rate H_S is fed character-by-character into a transmission line which may be live or dead. If the line is live when a character is transmitted then that character is received faithfully; if the line is dead then the receiver learns only that it is indeed dead. In shifting between its two states the line follows a Markov chain with constant transition probabilities, independent of the text being transmitted.

Show that the information rate of the source constituted by the received signal is $H_L + aH_S$, where H_L is the information rate of the Markov chain governing the functioning of the line and a is the equilibrium probability that the line is live. (Cambridge 1988)

11.   Consider a Bernoulli source (U_t) in which the individual character U_t can take value i with probability p_i, for $i = 1, 2, \ldots, m$. Let H denote the information rate. Let n_i be the number of times the

character i appears in the n-string $u^{(n)} = u_1 u_2 \ldots u_n$ of given length n, and let A_n be the smallest set of n-strings such that $P(U^{(n)} \in A_n) \geq 1 - \varepsilon$, where $0 < \varepsilon < 1$ is given. Show that each string in A_n satisfies the inequality

$$-\sum_i n_i \log p_i \leq nH + \sqrt{nK/\varepsilon},$$

where K is a constant not depending on n or ε. State (without proof) the analogous assertion for a simple Markov source. (Cambridge 1987)

12. Suppose a simple Markov source has transition probabilities (p_{jk}) with equilibrium distribution (w_j). Suppose a letter can be obliterated by noise (in which case one observes only the event 'erasure') with probability $\beta = 1 - \alpha$, independent of current or previous letter-values or previous noise. Show that the noise-corrupted source has information rate

$$-\alpha \log \alpha - \beta \log \beta - \alpha^2 \sum_j \sum_k \sum_{s=1}^{\infty} w_j \beta^{s-1} p_{jk}^{(s)} \log p_{jk}^{(s)},$$

where the $p_{jk}^{(s)}$ are the s-step transition probabilities of the original Markov chain. (Cambridge 1986)

13. A source $\mathbf{U}$ emits letters from the alphabet $F_m = \{1, \ldots, m\}$ as a Markov chain with transition probabilities $p_{ij} > 0$. The latter are such that the uniform distribution $w_i = 1/m$ $(i = 1, \ldots, m)$ is a stationary distribution. The initial letter emitted is uniformly distributed.

   The source becomes perturbed in that for each n with probability ε the nth emitted letter is replaced by a letter uniformly distributed over F_m, *independently* of the previously emitted letters. When this occurs the source continues *from the perturbed letter*. Thus if the perturbed n^{th} letter is i the $(n+1)^{\text{th}}$ letter is first chosen according to the probability distribution p_{ij} $(j = 1, \ldots, m)$, but may then be perturbed in the same way. Find the information rate $H(\mathbf{V})$ of the perturbed source $\mathbf{V}$. Show that, for some constant C,

$$H(\mathbf{V}) - H(\mathbf{U}) = C\varepsilon + O(\varepsilon) \qquad (\varepsilon \downarrow 0),$$

and evaluate C. (Cambridge Dipl. Stat. 1989)

14. Suppose a simple Markov source has transition probabilities p_{ij}. Each of the m states corresponds to a unique output symbol, and

every state can be reached from every other state. The *connection matrix* $A = (a_{jk})$ has $a_{jk} = 1$ if $p_{jk} > 0$, and $a_{jk} = 0$ if $p_{jk} = 0$. It can be shown that A has a real positive eigenvalue λ whose right column eigenvector v may be taken to have real positive elements v_j, and that (λ, v) is the only eigenvalue-eigenvector pair that is positive.

If $p_{jk} = a_{jk} v_k / (\lambda v_j)$, show that the information rate of the source is $\log \lambda$. Show also that this is the maximum rate for any source with the given connection matrix. (Cambridge Dipl. Stat. 1983)

15. Show that if $f : [0, 1] \to \mathbb{R}$ is a continuous function such that $\sum_1^m p_i f(p_i) = -c \sum_1^m p_i \log p_i$ for all $m > 1$ and for all $p_1, \ldots, p_m$ such that $p_i > 0$ and $\sum_1^m p_i = 1$, then $f(p) = -c \log p$.

16. Suppose the sequence $(X_t)_{t=\ldots,-2,-1,0,1,2,\ldots}$ is the output of a time-homogeneous Markov chain with a finite state-space and unique equilibrium-distribution. Quoting standard properties of entropy, show that

$$h(X_t | X_{t-1}) \le h(X_t | X_{t-2}) \le 2h(X_t | X_{t-1}).$$

The output of the source is modified by malfunction of the recording equipment: every third symbol is replaced by an unreadable splodge, $*$. Recording is started at a random time, so that

$$P(X_1 = *) = P(X_2 = *) = P(X_3 = *) = \tfrac{1}{3}.$$

Show that the information rate of the modified source is

$$\tfrac{1}{3}\left(h(X_t | X_{t-1}) + h(X_t | X_{t-2})\right),$$

and deduce that not more than one third of the information is lost. When is precisely $\frac{1}{3}$ lost?

For the case of a two-state Markov source with transition matrix

$$\begin{pmatrix} \tfrac{1}{3} & \tfrac{2}{3} \\ \tfrac{1}{2} & \tfrac{1}{2} \end{pmatrix},$$

find expressions for the information rate of the source, and of the source modified by illegibility of every third letter. (Cambridge 1984)

17. (a) Show that the 'entropy metric'

$$\Delta(X, Y) := h(X|Y) + h(Y|X)$$

is a pseudometric between random variables, i.e.

(i) $\Delta(X, Y) \ge 0, \quad \Delta(X, X) = 0$;

(ii) $\Delta(X, Y) = \Delta(Y, X)$;

(iii) $\Delta(X, Z) \leq \Delta(X, Y) + \Delta(Y, Z)$.

(b) Use Fano's inequality (Theorem 2.9.3) to show that Δ is continuous with respect to the metric $P(X \neq Y)$, that is, $P(X_n \neq Y_n) \to 0$ implies $\Delta(X_n, Y_n) \to 0$.

(c) Prove that $|h(X) - h(Y)| \leq \Delta(X, Y)$.

18. A random variable Y is distributed on the non-negative integers. Prove that the maximum entropy of Y, subject to $EY \leq c$, is $(c + 1)\log(c+1) - c\log c$, attained by a geometric distribution with mean c.

(Y has a *geometric distribution* if $P(Y = y) = p(1-p)^y$ for $y = 0$, 1, 2, ..., with $0 < p < 1$. It then has $EY = (1 - p)/p$.)

Hint: it suffices to find the Y of maximum entropy that has $EY = c$, because the maximized entropy is then increasing in c. Apply the Gibbs inequality with $p(\cdot)$ the distribution of Y and $q(\cdot)$ the above geometric distribution.

19. (a) Let f be a real-valued function on the range $\mathcal{X}$ of the r.v. X, and α an arbitrary real number. Prove that

$$h(X) \leq \alpha E f(X) + \log \sum_{x \in \mathcal{X}} e^{-\alpha f(x)}$$

with equality if and only if

$$P(X = x) = e^{-\alpha f(x)} \Big/ \sum_{y \in \mathcal{X}} e^{-\alpha f(y)} \qquad (x \in \mathcal{X}).$$

Hint: Apply the Gibbs inequality (Lemma 1.4.1) with $P(X = x)$ and $e^{-f(x)}$ in the roles of p_x and q_x respectively.

(b) Show that, for a positive integer-valued r.v. N,

$$h(N) < \log EN + \log e.$$

Hint: Put $f(n) := n$ and $\alpha := \log(EN/(EN - 1))$ in (a).

20. Consider a source for which U_1, U_2, ... is a stationary Markov chain with transition matrix $\left(\begin{smallmatrix} p & q \\ q & p \end{smallmatrix}\right)$ where $0 < p < 1$ and $q = 1 - p$. Show that, with ξ_n as in the First Coding Theorem (Theorem 2.3.2),

$$\xi_n = \frac{1}{n} - \frac{n-1}{n}\log p + \frac{1}{n}\left(\log\frac{p}{q}\right)\sum_{i=2}^{n} \mathbf{1}\{U_i \neq U_{i-1}\},$$

and that the r.v.s $\mathbf{1}\{U_i \neq U_{i-1}\}$ are independent Bernoulli r.v.s. Use the Strong Law of Large Numbers (§2.5) to obtain the a.s. convergence of ξ_n, i.e. the full Shannon-McMillan-Breiman theorem for this case.

21.   For each θ let $p(\theta) = (p_x(\theta))$ be a distribution on a finite set $\mathcal{X}$, where θ is a real parameter ranging over an open interval. Suppose that the probabilities $p_x(\theta)$ are positive and have continuous first derivatives with respect to θ. *Fisher's information* is defined to be

$$I(\theta) := \sum_{x \in \mathcal{X}} \frac{1}{p_x(\theta)} \left(\frac{\partial p_x(\theta)}{\partial \theta} \right)^2 .$$

Use l'Hôpital's rule to prove (2.13.3) (Kullback & Leibler (1951)).

RELIABLE TRANSMISSION

The agenda for this chapter is to establish what we want transmission channels to do and to find out if some typical ones can do it. We want a channel to transmit *reliably*, in a sense that we will define in §3.1 but that you should be able to guess by analogy with the 'reliable encoding' of Chapter 2. We will show in §§3.5–6 that two important types of channel, the binary symmetric and the memoryless Gaussian, can transmit reliably at positive rate, that is, they have non-zero *capacity*. That is at least as important as what their capacities actually turn out to be, which we will postpone the calculation of to Chapter 4. On the way to investigating particular channels we discuss how to decode possibly garbled messages, and that will involve a short treatment of some 'decision-theoretic' ideas. Then we meet *random coding*, thought up by Claude Shannon of the Bell Telephone System in the late 1940s, an extraordinary idea — so you should find it — and a powerful one, that underpins all of what can be proved about channel capacity.

3.1   Reliable transmission rates; channel capacity

Recall from Chapter 0 the transmission diagram

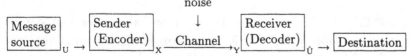

The source emits a random message $U^{(n)}$ of length n symbols. Treat it as a single word and encode it by a function γ_n, obtaining a coded message $X^{(n)} := \gamma_n(U^{(n)})$. Noise distorts this to $Y^{(n)}$, and

$$P(\text{receive } Y^{(n)} | X^{(n)} \text{ sent})$$

is prescribed (physically given). From $Y^{(n)}$, guess $U^{(n)}$ by $\hat{U}^{(n)}$, obtained by a *decoding rule*.

Let the effective set of possible values of $U^{(n)}$ be $\mathcal{U}_n$ and let $m = m_n := \#\mathcal{U}_n$. Recall from §2.2 that effectively $m = 2^{n(H+o(1))}$, where H is the source's information-rate. The code γ_n is to be chosen so that the set $\mathcal{X}_n := \{\gamma_n(u)\}_{u \in \mathcal{U}_n}$ of *codewords* or *waveforms* is within some (large) set $\mathcal{C}_n$ determined by the channel (e.g. all binary n-strings). We will assume $\#\mathcal{C}_n \geq \#\mathcal{U}_n$.

Definition. *A channel can transmit at a rate r bits/unit time reliably if given that $U^{(n)}$ can take $2^{n(r+o(1))}$ distinct* **equiprobable** *values, for each n a coding γ_n and a decoding rule can be found such that*

$$\beta(\gamma_n) := P(\hat{U}^{(n)} \neq U^{(n)}) \to 0 \qquad (n \to \infty).$$

So we are insisting that you can transmit at rate almost r in such a way that

$$P(U^{(n)} \text{ can be correctly inferred from } Y^{(n)}) \to 1.$$

Use of asymptotics means that we envisage long *segmented* codes. Also we ignore any storage problem or delay before our estimate $\hat{U}^{(n)}$ is released.

Definition. *The* capacity C *of the channel is the* **supremum of reliable transmission rates.**

These definitions and those of the previous chapter conceptually *decouple* the source from the channel. If the information rate H of the source does not exceed the channel capacity C then there is a number r such that $H \leq r \leq C$, and r is both a reliable encoding rate and a reliable transmission rate. (If $H = C$ the conclusions of Exercises 2.2.3 and 3.1.1 are needed here.) Thus *if $H \leq C$ then you can encode and transmit the source output reliably.*

The reasons for 'equiprobable' in the definition of reliable transmission rate are

(i)   the AEP (Asymptotic Equipartition Property, §2.4), and
(ii)   it is the worst case, as follows.

Lemma 3.1.1. *Fix n. Let $\varepsilon(p)$ be the error probability for source-message $U^{(n)}$ having probability distribution p, **minimized** over all codes and decoding rules. Then $\varepsilon(p)$ is worst (greatest) when $U^{(n)}$ is equidistributed over $\mathcal{U}_n$.*

Proof. For any coding $\gamma = \gamma_n$ that is non-injective, you can obviously improve the error probability by altering repeated values of γ to make it injective, as our assumption $\#\mathcal{C}_n \geq \#\mathcal{U}_n$ allows. So $\varepsilon(p)$ is the minimal error-probability over *injective* γ.

Suppose the possible values $u_1, \ldots, u_m$ of $U^{(n)}$ have respective probabilities $\pi_1, \ldots, \pi_m$, and the injective coding γ gives them distinct codewords $x_1, \ldots, x_m$. Let $\beta_j := P(x_j \text{ not inferred} \mid x_j \text{ sent})$. Then

$$P(\text{error}) = \sum_{j=1}^{m} \pi_j \beta_j,$$

since you decode from x_j to u_j without error.

If you permute the allocation of codewords, i.e. code u_j by $x_{\sigma(j)}$ where σ is a permutation of $\{1, \ldots, m\}$, you get probability of error $\beta(\sigma) = \sum_1^m \pi_j \beta_{\sigma(j)}$. If the source words are equiprobable this becomes

$$\sum_1^m \frac{1}{m} \beta_{\sigma(j)} = \frac{1}{m} \sum_1^m \beta_j =: \bar{\beta},$$

the same for all σ.

Claim: for any π there exists σ_π such that $\beta(\sigma_\pi) \leq \bar{\beta}$.

That is, *for any source distribution π you can re-code to bring the error probability to at least as good a value as for the equiprobable source-distribution.*

To see this, let Σ be a *random* permutation, equally likely to be any of the elements of S_m, the set of all permutations of $\{1, \ldots, m\}$. Then

$$\min_{\sigma \in S_m} \beta(\sigma) \leq E\beta(\Sigma)$$

$$= E \sum_1^m \pi_j \beta_{\Sigma(j)}$$

$$= \sum_1^m \pi_j E\beta_{\Sigma(j)}$$

$$= \sum_1^m \pi_j (\beta_1 + \cdots + \beta_m)/m$$

$$= (\beta_1 + \cdots + \beta_m)/m = \bar{\beta}. \qquad \square$$

Obviously, for non-equiprobable π, the allocation that gives the smallest error-probability is

$$\text{large } \beta \text{ with small } \pi.$$

That is, if $\pi_1 \leq \cdots \leq \pi_m$ then you want $\beta_1 \geq \cdots \geq \beta_m$.

Exercises

1. Show that the supremum in the definition of channel capacity C is attained, so that C is itself a reliable transmission rate.
    Hint: adapt your solution of Exercise 2.2.3.

2. Prove the claim in the proof of Lemma 3.1.1, that for any source-distribution π you can re-code to bring the error probability at least

as low as for the equiprobable source-distribution, in the following alternative way. Let the π_j be indexed so that $\pi_1 \leq \cdots \leq \pi_m$ and let ρ be the permutation such that $\beta_{\rho(1)} \geq \cdots \beta_{\rho(m)}$. Show that unless the π_j are all equal there exist $r < s$ such that $\pi_r < 1/m < \pi_s$, and $\pi_j = 1/m$ for all j with $r < j < s$. Show that increasing π_r by ε, and decreasing π_s by ε, such that the closer of them to $1/m$ becomes $1/m$, you increase $\sum_1^m \pi_j \beta_{\rho(j)}$. Repeating the process if necessary, you get to $\bar{\beta}$ in at most $m - 1$ steps.

3.2   Decoding: receiver optimization

Send a coded message block $X^{(n)} = \gamma(U^{(n)})$; the channel emits $Y^{(n)}$. *The channel is specified by $P(Y^{(n)} = y^{(n)}|X^{(n)} = x^{(n)})$, the conditional probability distribution of $Y^{(n)}$ given $X^{(n)} = x^{(n)}$.*

(More generally you might need to specify $P(Y^{(n)} = y^{(n)}|X^{(n)} = x^{(n)}, S = s)$ where S is the *state* of the channel — e.g. alive/dead, upper-case/lower-case — which can change over time. But we will spare you that.)

The set of possible received messages $\mathcal{Y}$, say, is in general larger than the set of possible transmitted messages, as corruption will give 'illegal' words. From a received y you guess x. The set of y that you decode as a particular x is called a *decoding set*. It is natural to have each legal message x belonging to its own decoding set, so the mental picture is something like Fig. 3.2.1.

To be considered are

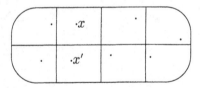

Figure 3.2.1. Decoding sets

- choice of decoding sets,
- criteria for choice.

A little decision theory

The left-hand column below gives some general notions from the subject known as decision theory, and the right-hand column gives the relevant specialization in each case to our set-up.

Suppose there is an unobservable x, the 'state of nature', the value of an $\mathcal{X}$-valued r.v. X.

(*x is the message **sent**, the value of the random message X. We omit the sub- or super-script n.*)

We are to take an 'action' a from $\mathcal{A}$, the set of possible actions.

(*a is a guess at x, so $\mathcal{A} = \mathcal{X}$, the set of legal messages.*)

A loss $l(x, a)$ results if action a is taken when x is the true state of nature.

(A loss $l(x, a)$ results if we guess that the message sent is a when it is in fact x.)

Observe $y \in \mathcal{Y}$, the value of a r.v. Y.

(y is the message **received**.)

We need to develop the above ideas a little way in general terms, but if you keep in mind the interpretations given then the following will make sense.

Consider a *decision rule* $d : \mathcal{Y} \to \mathcal{A}$, so that

$$y \text{ received } \implies \text{ action } d(y) \in \mathcal{A}.$$

The loss becomes a random variable $l(X, d(Y))$.

Criterion. Choose d to minimize *expected loss* $\bar{l}(d) := El(X, d(Y))$.

The discrete case

You have a known prior distribution $\pi_x := P(X = x)$, for $x \in \mathcal{X}$. There is a known conditional distribution $f(y|x) := P(Y = y|X = x)$, where $x \in \mathcal{X}$, $y \in \mathcal{Y}$ and $\mathcal{X}$ and $\mathcal{Y}$ are *countable*. Then

$$\bar{l}(d) = \sum_x \sum_y l(x, d(y)) \pi_x f(y|x).$$

Theorem 3.2.1(D). *For each y let $\hat{a}(y)$ be the value of a that minimizes $\sum_x l(x, a)\pi_x f(y|x)$. Then the expected loss $\bar{l}(d)$ is least when $d = \hat{a}$.*

Proof.

$$\bar{l}(d) = \sum_y \sum_x l(x, d(y)) \pi_x f(y|x) \geq \sum_y \sum_x l(x, \hat{a}(y)) \pi_x f(y|x) = \bar{l}(\hat{a}). \quad \square$$

The continuous case

You have a prior *density* $\pi(x)$ for the real-valued r.v. X, and there is a conditional *density* $f(y|x)$ for Y given $X = x$. So

$$\bar{l}(d) = \iint l(x, d(y)) \pi(x) f(y|x) dy \, dx.$$

Theorem 3.2.2(C). *For each y let $\hat{a}(y)$ be the value of a that minimizes $\int l(x, a)\pi(x)f(y|x)dx$. Then $\bar{l}(d)$ is least when $d = \hat{a}$.*

Proof. The same, with sums replaced by integrals. $\quad \square$

The theorem reduces minimization with respect to functions d to minimization with respect to points a — which should be an easier task.

Now Bayes's rule gives the *posterior* distribution for X given Y:

$$P(X = x|Y = y) = \frac{\pi_x f(y|x)}{\sum_{z \in \mathcal{X}} \pi_z f(y|z)} \qquad \text{(discrete)}$$

$$f_{X|Y}(x|y) = \frac{\pi(x) f(y|x)}{\int_{\mathcal{X}} \pi(z) f(y|z) dz} \qquad \text{(continuous)}.$$

Moral (Corollary 3.2.3). *The optimum d is obtained by minimizing the posterior expected loss* $E(l(X,a)|Y = y)$.

(For the latter is

$$\sum_x l(x,a) \frac{\pi_x f(y|x)}{\sum_z \pi_z f(y|z)}, \text{ respectively } \int l(x,a) \frac{\pi(x) f(y|x)}{\int \pi(z) f(y|z) \, dz} dx,$$

and note that the denominator in each case is constant with respect to a.)

ML decoding

We now specialize the decision theory to the channel case. So x is now a message value, belonging to $\mathcal{X}$, the set of possible messages. And $X = X^{(n)} = \gamma(U^{(n)})$ is a coded message, while a is to be a guess at what x is, so had better belong to $\mathcal{X}$. That is, $\mathcal{A} = \mathcal{X}$. (For the moment assume γ injective, so there is error-free decoding of x to u.) Take

$$l(x,a) := \mathbf{1}\{x \neq a\} = \begin{cases} 0 & \text{if } a = x \\ 1 & \text{if not} \end{cases}.$$

Considering the discrete case, you find that $\hat{a}(y)$ is the value of $a \in \mathcal{X}$ that minimizes

$$\sum_x \mathbf{1}\{x \neq a\} \pi_x f(y|x) = \sum_x \pi_x f(y|x) - \sum_x \mathbf{1}\{x = a\} \pi_x f(y|x)$$

$$= \sum_x \pi_x f(y|x) - \pi_a f(y|a).$$

So $\hat{a}(y)$ is the value of a that maximizes $\pi_a f(y|a)$. Rename it $\hat{x}(y)$.

Definitions. *The* ideal-observer *or* minimum-error *decoder decodes y by $\hat{x}(y)$, the value of x that maximizes $\pi_x f(y|x)$. That is, $\hat{x}(y)$ is the x that maximizes $P(X = x|Y = y) = \pi_x f(y|x) / \sum_{z \in \mathcal{X}} \pi_z f(y|z)$.*

The ML *or* maximum-likelihood *decoder decodes y by $\tilde{x}(y)$, the value of x that maximizes $f(y|x) = P(Y = y|X = x)$, the (conditional) likelihood.*

When the source messages are equiprobable, if γ is injective obviously the coded messages are equiprobable too, so $\pi_x = 1/m$ for all $x \in \mathcal{X}$, where $\#\mathcal{X} = m$.

When the coded messages are equiprobable and the code is injective the ML and ideal-observer decoders coincide.

In using the ML decoder, error can occur either because it chooses the wrong x, or because γ is non-injective and the value u of the source message $U = U^{(n)}$ is such that $\gamma(u) = \gamma(u')$ for some other u'.

Lemma 3.2.4. *For equidistributed source-messages in $\mathcal{U}$, where $\#\mathcal{U} = m$, and for the ML decoder, the probability $\beta(\gamma)$ of error when using coding γ satisfies*

$$\beta(\gamma) \le \frac{1}{m} \sum_u \sum_{\{u':u' \ne u\}} P(f(Y|\gamma(u')) \ge f(Y|\gamma(u)) \mid U = u).$$

Proof. If the source emits u you get

- an error if $f(Y|\gamma(u')) > f(Y|\gamma(u))$ for some $u' \ne u$,
- possibly an error if $f(Y|\gamma(u')) = f(Y|\gamma(u))$ for some $u' \ne u$,
- no error if $f(Y|\gamma(u')) < f(Y|\gamma(u))$ for all $u' \ne u$,

the middle alternative covering cases when $\gamma(u) = \gamma(u')$. So

$$P(\text{error}|U = u) \le P(f(Y|\gamma(u')) \ge f(Y|\gamma(u)) \text{ for some } u' \ne u|U = u)$$
$$\le \sum_{\{u':u' \ne u\}} P(f(Y|\gamma(u')) \ge f(Y|\gamma(u))|U = u)$$

by Boole's inequality (i.e. $P(\cup A_k) \le \sum A_k$, for any finite or countable collection of events (A_k)). Multiply by $\frac{1}{m} = P(U = u)$ and add over u, hence the result. $\qquad \square$

3.3 Random coding

We have already been using the following notion of a (channel) code:

Definition. *A coding or code is a map $\gamma : \mathcal{U}_n \to \mathcal{C}_n$.*

This is the same definition as those we gave for noiseless coding in §1.1 and informally for source coding in §2.13, but the notation for the code and its domain and range has changed. That is appropriate because these different varieties of coding pertain to messages at different stages of their processing, and can all be used in series if needed, with the range of one code-function serving as the domain for the next.

Consider a code γ as defined above. Identifying it with the set $\mathcal{X} = \mathcal{X}_n := \{\gamma(u) : u \in \mathcal{U}_n\}$ of *codewords*, you can consider γ as an element of $\mathcal{C}_n^m$.

Consider a *random* such map Γ. That is, Γ is to be a random element of $\mathcal{C}_n^m$, *independent* of the source $\mathbf{U}$, with some distribution

$$P(\Gamma = \gamma) \qquad (\gamma \in \mathcal{C}_n^m).$$

Amazingly, this can achieve a good coding, just with Γ equidistributed over $\mathcal{C}_n^m$. For example, suppose there are $m = 2^{10} = 1024$ possible messages, to be coded by binary 20-strings, so $\#\mathcal{C}_n = 2^{20} \simeq 10^6$. Then there are $(2^{20})^m = 2^{20480}$ ways to map $\mathcal{U}_n$ into $\mathcal{C}_n$, i.e. to choose 1024 codewords from the 10^6 available. Doing so at random you are *very likely* to get *highly distinct* codewords in $\mathcal{C}_n$, so that, intuitively, minor corruptions can be easily put right, and decoding done without error.

More formally, reasons for considering random codings are

- an existence theorem for good codes;
- you can calculate over averages of equidistributed random codes (more symmetric calculations);
- discrete optimization over $x_1, \ldots, x_m$ is replaced by easier continuous optimization over probability distributions.

As above, $\beta(\gamma)$ denotes the probability of error, for equiprobable source-words, when using coding γ and the ML decoder. So for a random coding Γ the expected probability of error is

$$\bar{\beta} := E\beta(\Gamma) = \sum_\gamma \beta(\gamma) P(\Gamma = \gamma).$$

Lemma 3.3.1.

(i)   *There exists γ with $\beta(\gamma) \le \bar{\beta}$.*

(ii) $P(\beta(\Gamma) \le \bar{\beta}/(1-\rho)) \ge \rho$, *for $0 \le \rho \le 1$.*

Proof. (i) is immediate (not everything can be above the average).

For (ii) use *Markov's inequality:* for any non-negative r.v. X,

$$P(X \ge a) \le a^{-1} EX \qquad (a > 0) \tag{1}$$

(see Exercise 1). So

$$P(\beta(\Gamma) \le a) \ge 1 - \bar{\beta}/a \qquad (a > 0);$$

put $a := \bar{\beta}/(1-\rho)$. $\square$

Exercise

1.   Prove Markov's inequality (3.3.1).

        Hint: note that $X \ge a\mathbf{1}\{X \ge a\}$ and take expectations.

3.4   Introduction to channels

Recall that the channel is specified by $P(Y^{(n)} = y^{(n)}|X^{(n)} = x^{(n)})$ for all arguments n, $x^{(n)}$, $y^{(n)}$, where $X^{(n)}$ and $Y^{(n)}$ are the random n-strings $X_1 X_2 \ldots X_n$ and $Y_1 Y_2 \ldots Y_n$ respectively.

The channel is *discrete* if X_t, Y_t have finitely many values (the same for all t).

The channel is *lossless*, or *perfect*, if X_t can be inferred with certainty from Y_t.

The channel is *useless* if Y_t is independent of X_t, whatever the distribution of X_t.

The channel is *deterministic* if the input determines the output, with certainty.

The channel is *memoryless* if

$$P(Y^{(n)} = y^{(n)}|X^{(n)} = x^{(n)}) = \prod_{t=1}^{n} P(Y_t = y_t|X_t = x_t)$$

for all n, $x^{(n)}$, $y^{(n)}$. (Recall that $y^{(n)}$ is the n-string $y_1 y_2 \ldots y_n$ and $x^{(n)}$ is $x_1 x_2 \ldots x_n$.) Thus the memoryless channel is one which, given the X_t, decides the Y_t independently of each other.

For a discrete memoryless channel (DMC) the *channel matrix* is $P = (p_{jk})$ where $p_{jk} = P(Y_t = k|X_t = j)$. (So everything is assumed stationary in time. But this matrix is *not* a Markov-chain transition matrix: it need not even be square.)

A DMC is *symmetric* if the rows of the channel matrix are permutations of each other, and so are the columns. Examples of matrices of such channels are

$$\begin{pmatrix} \frac{1}{2} & \frac{1}{3} & \frac{1}{6} \\ \frac{1}{6} & \frac{1}{2} & \frac{1}{3} \\ \frac{1}{3} & \frac{1}{6} & \frac{1}{2} \end{pmatrix}, \qquad \begin{pmatrix} \frac{1}{3} & \frac{1}{3} & \frac{1}{6} & \frac{1}{6} \\ \frac{1}{6} & \frac{1}{6} & \frac{1}{3} & \frac{1}{3} \end{pmatrix}.$$

A channel is *binary* if the input and output alphabets have two elements.

So the *binary symmetric channel* (BSC) has input and output alphabets 0, 1 and channel matrix $P = \begin{pmatrix} q & p \\ p & q \end{pmatrix}$. Here $p = 1 - q$ is the probability of either kind of error, and the channel is lossless if $p = 0$ (or 1), useless if $p = \frac{1}{2} = q$. The BSC was formerly believed to have what we would now call capacity 0, if $0 < p < 1$ (see the

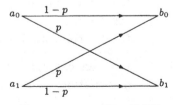

Figure 3.4.1. The BSC

end of §3.5).

An obvious way to picture the BSC is as shown in Fig. 3.4.1.

The *binary erasure channel* is rather similar. In it, the transmitted symbol can be corrupted into an illegible splodge '*', as in the left-hand diagram in Fig. 3.4.2. The right-hand diagram is the *binary error-and-erasure channel*.

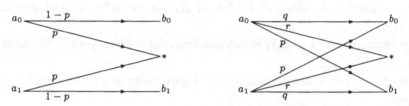

Figure 3.4.2. Binary (error-and-) erasure channels

Exercises

1.   Prove that if the matrix P of a symmetric channel is $m \times n$ then it has at most (m, n) distinct elements, where (m, n) denotes the greatest common divisor of m and n.

        Hint: counting by rows, there are $r_i m$ elements of type i, where r_i is the number of elements of type i in any one row. Counting by columns, there are $c_i n$ elements of type i. Thus $r_i m = c_i n$. For r_i and c_i to be minimal (and hence maximize the number of distinct elements), $(r_i, c_i) = 1$. Then $r_i = n/(m, n)$ and $c_i = m/(m, n)$. Thus there are at least $r_i m = mn/(m, n)$ of type i for each i, and so at most $mn/(r_i m) = (m, n)$ distinct elements.

2.   Use the result of Exercise 1 to find the general form of a channel matrix P for a symmetric channel when P is (i) 6×4, (ii) 9×6.

3.   Show that a DMC is
        (a) useless iff all the rows of the channel matrix are the same;
        (b) lossless iff each column of the channel matrix has only one non-zero entry;
        (c) deterministic iff all entries of the channel matrix are 0 or 1.

4.   Give an example of a non-useless DMC, and a particular input-distribution, such that Y_t is independent of X_t.

5.   A message to be sent to you through a BSC is either 1000 or 0110 or 0001 or 1111, these having probabilities $\frac{1}{3}$, $\frac{1}{3}$, $\frac{1}{6}$, $\frac{1}{6}$ respectively. All this is known to you, as is the error probability $p = \frac{1}{10}$ of the

BSC. Suppose you receive 1001. Which of the four messages would you decode it as, using
  (i) ideal-observer decoding,
  (ii) ML decoding?

3.5 Reliable transmission through the BSC

Definitions. *Between n-strings $x = x_1 x_2 \ldots x_n$, $y = y_1 y_2 \ldots y_n$ the Hamming distance is*

$$\rho(x, y) := \sum_{t=1}^{n} 1\{x_t \neq y_t\} = \text{ no. of places where } x_t, y_t \text{ differ.}$$

Hamming distance is indeed a metric (Exercise 2).

When the individual symbols in the strings are real numbers we can define the L^1 metric $\rho_1(x, y) := \sum_{t=1}^{n} |x_t - y_t|$. This differs from the Hamming metric in general, but coincides with it for binary strings.

We can use the Hamming metric to analyze the BSC. First,

$$P(Y^{(n)} = y | X^{(n)} = x) = p^{\rho(x,y)} q^{n - \rho(x,y)} = q^n (p/q)^{\rho(x,y)}. \tag{1}$$

The transmitter sends any binary n-string from the set $\mathcal{X} \subset \{0, 1\}^n$, where $\# \mathcal{X} = m$ and $m = 2^{n(r + o(1))}$. The receiver can receive *any* binary n-string y. Use the ML decoder (§3.2) to decode it. In a binary channel there is only one wrong symbol resulting from a symbol being transmitted incorrectly. Hence the ML rule simplifies, as (1) confirms, to the following.

The ML decoding rule: decide for a codeword x from $\mathcal{X}$

$$\begin{cases} \text{minimizing } \rho(x, y) & \text{if } p < \frac{1}{2}, \\ \text{maximizing } \rho(x, y) & \text{if } p > \frac{1}{2}. \end{cases}$$

Lemma 3.5.1. *For equidistributed source-messages in $\mathcal{U}$, where $\#\mathcal{U} = m$, and for the ML decoder, the BSC has*

$$\beta(\gamma) \leq \frac{1}{m} \sum_u \sum_{\{u' : u' \neq u\}} \theta^{\rho(\gamma(u), \gamma(u'))} \qquad \text{where } \theta := \sqrt{4pq}.$$

(Note $\theta \leq 1$, with equality iff $p = \frac{1}{2}$.)

Proof. Assume $p < \frac{1}{2}$. With $f(y|x) := P(Y^{(n)} = y | X^{(n)} = x)$ you have for any x, x' (possibly equal) that

$$P(f(Y|x') \geq f(Y|x) \mid X = x) = P(\rho(x', Y) \leq \rho(x, Y) \mid X = x)$$
$$= P(S \leq 0 | X = x)$$
$$\text{(where } S := \rho(x', Y) - \rho(x, Y))$$

$$\leq E(z^S|X = x)$$

for any $z \in (0, 1]$. The last inequality is because $\mathbf{1}\{s \leq 0\} \leq z^s$ for all s, as Fig. 3.5.1 illustrates.

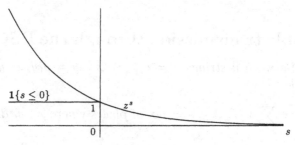

$$\text{Figure 3.5.1}$$

Write $x = x_1 x_2 \ldots x_n$, etc., then $S = \sum_1^n \Delta_t$ where $\Delta_t := \mathbf{1}\{Y_t \neq x'_t\} - \mathbf{1}\{Y_t \neq x_t\}$ so that

$$\Delta_t = \begin{cases} 0 & \text{if } x_t = x'_t, \\ 1 & \text{if } x_t \neq x'_t \text{ and } Y_t = x_t \quad \text{(with probability } q, \text{ given } X = x). \\ -1 & \text{if } x_t \neq x'_t \text{ and } Y_t = x'_t \quad \text{(with probability } p, \text{ given } X = x). \end{cases}$$

(Cases of coincident codewords, $x = x'$, are covered by the first alternative here.) Thus

$$E(z^S|X = x) = E\left(\prod_{t=1}^n z^{\Delta_t} \,\Big|\, X = x\right)$$

$$= \prod_1^n E(z^{\Delta_t} \mid X = x) \qquad \text{(since memoryless)}$$

$$= \prod_{\{t: x_t = x'_t\}} z^0 \prod_{\{t: x_t \neq x'_t\}} (pz^{-1} + qz)$$

$$= (pz^{-1} + qz)^{\rho(x,x')}.$$

Therefore

$$P(f(Y|x') \geq f(Y|x) \mid X = x) \leq (pz^{-1} + qz)^{\rho(x,x')}.$$

Give z its minimizing value $\sqrt{p/q}$, so

$$P(f(Y|x') \geq f(Y|x) \mid X = x) \leq \theta^{\rho(x,x')}.$$

For distinct u, u' in $\mathcal{U}$, set $x := \gamma(u)$, $x' := \gamma(u')$, then the conditional distribution of Y given $U = u$ is the same as that given $\gamma(U) = x$, so

$$P(f(Y|\gamma(u')) \geq f(Y|\gamma(u)) \mid U = u) \leq \theta^{\rho(\gamma(u),\gamma(u'))}.$$

Insert this into Lemma 3.2.4, hence the claimed bound.

The above proof works also for $p > \frac{1}{2}$, since the ML rule flips over. □

Let the set of possible source-messages $\mathcal{U}_n$ be $\{u_1, \ldots, u_m\}$. Suppose a random coding Γ is applied, then the random set of codewords is $\Gamma(u_1), \ldots,$ $\Gamma(u_m)$, and $\Gamma(u_i)$ is the random binary string $W_{i1}W_{i2}\ldots W_{in}$, say. The set $\mathcal{C}_n$ is to be $\{0,1\}^n$, the set of all binary n-strings. Now take

Γ equidistributed over $\mathcal{C}_n^m$, independent of U;
—equivalently $\Gamma(u_1)$, $\Gamma(u_2)$, $\ldots$, $\Gamma(u_m)$ independently equidistributed over $\mathcal{C}_n$;
—equivalently W_{it}, for $i = 1, \ldots, m$ and $t = 1, \ldots, n$, independently equidistributed over $\{0,1\}$.

Lemma 3.5.2. *For the random coding Γ, equidistributed over $\mathcal{C}_n^m$, and for equiprobable source-messages and ML decoding, the expected probability of error satisfies*

$$\bar{\beta}_n := E\beta(\Gamma) \le \left(2^{r+o(1)}\left(\frac{1+\theta}{2}\right)\right)^n.$$

Proof. Pick $u \ne u'$ in $\mathcal{U}$. Then the probability that $\Gamma(u)$ and $\Gamma(u')$ differ in (exactly) j places is $\binom{n}{j}\left(\frac{1}{2}\right)^j\left(\frac{1}{2}\right)^{n-j}$, so

$$E\theta^{\rho(\Gamma(u),\Gamma(u'))} = \sum_{j=0}^{n}\binom{n}{j}\left(\frac{1}{2}\right)^j\left(\frac{1}{2}\right)^{n-j}\theta^j = \left(\frac{1}{2}+\frac{\theta}{2}\right)^n.$$

Therefore

$$E\beta(\Gamma) \le (m-1)\left(\frac{1+\theta}{2}\right)^n = \left(2^{r+o(1)}\left(\frac{1+\theta}{2}\right)\right)^n. \qquad \square$$

Theorem 3.5.3. *For the BSC, with error probability $p = 1 - q$,*

$$C \ge 1 - \log(1 + 2\sqrt{pq}).$$

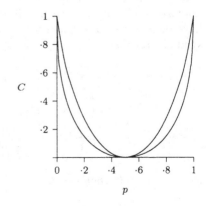

Figure 3.5.2

Proof. If $r < \log(2/(1+\theta))$ then $\bar{\beta}_n \to 0$ as $n \to \infty$, by Lemma 3.5.2. By Lemma 3.3.1(i), for each n there exists a non-random coding with error probability at most $\bar{\beta}_n$, so reliable transmission is possible at rate r. Therefore

$$C \ge \log\left(\frac{2}{1+\theta}\right)$$
$$= 1 - \log(1 + 2\sqrt{pq}). \qquad \square$$

The true C is $1 + p \log p + q \log q$, as we

shall see later. It is the upper curve in Fig. 3.5.2, while the bound we have just established is the lower curve.

As the length of the received block is the same, n symbols, as the length of the transmitted block, the units of r are now bits per source-symbol. If we send the coded block $X^{(n)}$ in time n this becomes bits per unit time as before.

In the above proofs we have let the block-length n tend to infinity. What we have shown is thus that for large n we can find a code to transmit a whole n-block with small probability of *any* error therein, thus *correcting* channel errors.

Random coding is here a device to get existence. We calculated means over the whole 'ensemble' of possible codes and used the fact that there must exist some member no worse than the mean. That is our 'good code', at which reliable transmission is possible at positive rate. We have felt it worthwhile to establish this in a relatively direct way, even though the lower bound we have established for C will become redundant when we calculate C explicitly in the next chapter.

It is, though, worth emphasizing that we have so far obtained only an existence proof: *it gives no construction of a single 'good code'*. In Chapters 5 and 6 we shall cover explicit construction of codes with 'good' error-correction properties, as well as further relationships between block-length n and the rate at which errors can be corrected.

Example 3.5.4. This will show that *naïve coding does not give a positive reliable transmission rate*. Consider the BSC with error probability $p < \frac{1}{2}$. Suppose the transmitter repeats each symbol $2m + 1$ times. This is called a *repetition code* and you will meet it again in Chapter 5 (Example 5.1.4). You decode it by the 'majority' rule: if more than half of the $2m + 1$ symbols are 1, decide a 1 must have been intended, otherwise a 0. The probability that you decide a transmitted symbol wrongly is

$$
\pi_m := \sum_{k=m+1}^{2m+1} \binom{2m+1}{k} p^k (1-p)^{2m-k+1}
$$

$$
= P(T \geq m+1) \qquad \text{where } T \text{ is Binomial}(2m+1, p)
$$

$$
= P\left(\frac{T - (2m+1)p}{\sqrt{(2m+1)pq}} \geq \frac{m(1-2p)+q}{\sqrt{(2m+1)pq}} \right).
$$

Pick any $K > 0$, then because $p < \frac{1}{2}$ this is for large m at most

$$
P\left(\frac{T - (2m+1)p}{\sqrt{(2m+1)pq}} \geq K \right),
$$

which tends to $1 - \Phi(K)$ as $m \to \infty$, by the Central-limit Theorem (§2.5). So $\pi_m \to 0$ as $m \to \infty$, since $1 - \Phi(K)$ can be made as small as you wish.

Consider a large block of n received symbols made up of $l = n/(2m+1)$ such groups, so you have a transmission rate of $r = 1/(2m+1)$. The probability of error somewhere in the block is

$$1 - (1 - \pi_m)^l,$$

which tends to 0 iff $l\pi_m \to 0$, that is, $n\pi_m/(2m+1) \to 0$. So m must change with n and tend to ∞. So $r \to 0$.

Exercises

1. Consider a BSC with error probability ε. The only codewords used are 000 and 111. Show that under ML decoding the probability of error is $3\varepsilon^2 - 2\varepsilon^3$.

2. Consider Hamming distance ρ on the set F^n of n-strings of elements of some alphabet F. Prove that it is a metric on F^n, that is,
   (i) $\rho(x, y) \geq 0$, with equality iff $x = y$;
   (ii) $\rho(x, y) = \rho(y, x)$ (symmetry);
   (iii) $\rho(x, y) \leq \rho(x, z) + \rho(z, y)$ (triangle inequality).
   On $\mathbb{R}^n$, does the Hamming metric give rise to the same topology (same open sets) as the L^1 metric?

3. Let X and Y be independently distributed over the set of binary n-strings. Show that the Hamming distance between them has the Binomial$(n, \frac{1}{2})$ distribution.
   (a) Deduce that the probability that X and Y differ in at least 2 digits is $1 - (n+1)/2^n$. Find how large n must be in order that this probability should exceed 0·95.
   (b) Show that the average Hamming distance between binary n-strings x and y, over all choices of x and y, is $\frac{1}{2}n$.

4. Consider a BSC with error probability p satisfying $0 < p < \frac{1}{2}$, and suppose you transmit through it at fixed rate r satisfying $0 < r < 1 - \log(1 + 2\sqrt{p(1-p)})$, encoding n-blocks of source letters by a code for which with a suitable decoder there is *minimal* probability $\beta_{n,r}$ of any error in the n-block. Deduce from Lemma 3.5.2 that $\beta_{n,r} \leq e^{-n\varepsilon}$ where $\varepsilon = 1 - \log(1 + 2\sqrt{p(1-p)}) - r > 0$.

3.6 Reliable transmission through the memoryless Gaussian channel

Recollections of normality

As we mentioned in §2.5, the random variable Z is *standard normal* (or *standard Gaussian*) if it has probability density

$$\phi(z) := (2\pi)^{-\frac{1}{2}}e^{-\frac{1}{2}z^2} \qquad (z \in \mathbb{R}),$$

and thus (cumulative) distribution function

$$\Phi(t) := P(Z \leq t) = \frac{1}{\sqrt{2\pi}}\int_{-\infty}^{t} e^{-\frac{1}{2}z^2}dz.$$

The distribution is denoted $N(0,1)$. For the tail integral $1-\Phi(t) = P(Z > t)$ you have the following useful inequality.

Lemma 3.6.1. $\qquad 1 - \Phi(t) \leq e^{-\frac{1}{2}t^2} \qquad (t \geq 0).$

Proof. For $\theta \geq 0$,

$$\begin{aligned}
P(Z > t) &= \frac{1}{\sqrt{2\pi}}\int_{t}^{\infty} e^{-\frac{1}{2}z^2}dz \\
&\leq \frac{1}{\sqrt{2\pi}}\int_{-\infty}^{\infty} e^{\theta(z-t)}e^{-\frac{1}{2}z^2}dz \qquad \text{(see Fig. 3.6.1)} \\
&= e^{\frac{1}{2}\theta^2-\theta t}\frac{1}{\sqrt{2\pi}}\int_{-\infty}^{\infty} e^{-\frac{1}{2}(z-\theta)^2}dz = e^{\frac{1}{2}\theta^2-\theta t}.
\end{aligned}$$

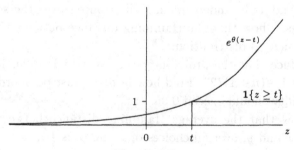

Figure 3.6.1

Set $\theta := t$. $\qquad\qquad\qquad\qquad\qquad\qquad\qquad\qquad\qquad\qquad\qquad$ □

In fact

$$1 - \Phi(t) \sim \phi(t)/t \qquad (t \to \infty), \tag{1}$$

a result known as *Mills's ratio*; see Exercise 2. The notation $f(t) \sim g(t)$ means that $f(t)/g(t) \to 1$.

The general normal (Gaussian) distribution $N(\mu, \sigma^2)$ on $\mathbb{R}$ is that of a random variable X, say, with density

$$f(x) = (2\pi\sigma^2)^{-\frac{1}{2}} e^{-\frac{1}{2}(x-\mu)^2/\sigma^2} \qquad (x \in \mathbb{R}).$$

It has mean μ and variance σ^2. Also $(X - \mu)/\sigma$ is then $N(0,1)$, hence $P(X > t) = 1 - \Phi((t - \mu)/\sigma)$ which you can bound as above.

The memoryless Gaussian channel (MGC)

As usual, the message is encoded as an n-string $X^{(n)} = X_1 X_2 \ldots X_n$ and transmitted down the channel. You receive an n-string $Y^{(n)} = Y_1 Y_2 \ldots Y_n$, where

$$Y_t = X_t + \varepsilon_t \tag{2}$$

and the ε_t are independent $N(0, v)$, independent of $(X_1, X_2, \ldots)$.

The symbols X_t thus have to be real numbers, though they can still be restricted to some finite alphabet if we wish. However under the model the Y_t can take all real values and thus we lose the discrete value-space that we have used to good effect earlier. Of course there is plenty of analytic machinery to take its place.

We note also that 'thermal' noise in radio devices is additive Gaussian, so this model is of key importance physically.

As before, $U = U^{(n)}$ is equidistributed over $\mathcal{U}_n$, where $\#\mathcal{U}_n = m = 2^{n(r+o(1))}$, and $\gamma : \mathcal{U}_n \to \mathcal{C}_n$, so $X^{(n)} := \gamma(U^{(n)})$.

What is new is that the random n-string $Y^{(n)} = Y_1 Y_2 \ldots Y_n$ has a continuous distribution; in fact, given $X^{(n)} = x = x_1 x_2 \ldots x_n$, the Y_t are *independent* $N(x_t, v)$. So, given $X^{(n)} = x$, the density of $Y^{(n)}$ over $\mathbb{R}^n$, that is, the joint density of $Y_1, \ldots, Y_n$, is

$$f(y|x) = \prod_{t=1}^{n} \frac{1}{\sqrt{2\pi v}} e^{-\frac{1}{2}(y_t - x_t)^2/v} = (2\pi v)^{-\frac{n}{2}} e^{-\frac{1}{2}|x-y|^2/v},$$

where y is the string $y_1 y_2 \ldots y_n$ and $|x - y| := \sqrt{\sum_1^n (x_t - y_t)^2}$.

This we now consider as the *likelihood*, so the ML decoder decodes y by $\tilde{x}(y)$, the value of x that maximizes $f(y|x)$.

When the source messages are equiprobable, and the coding γ is injective, this coincides with the ideal-observer decoder, by the same reasoning as before (§3.2).

Also Lemma 3.5.1 carries over to this setting, with unaltered proof.

ML decoding rule

From the above evaluation of $f(y|x)$, the ML decoding rule is to **decide for the codeword x from $\mathcal{X}$ that minimizes $|y - x|$.**

Thinking of x, y temporarily as columns $(x_1, \ldots, x_n)^\top$, $(y_1, \ldots, y_n)^\top$, you have $|y - x|^2 = |y|^2 - 2x^\top y + |x|^2$. If all codewords x have the same 'energy' $|x|^2 = \sum_1^n x_t^2$ then you choose the codeword x *maximizing* $x^\top y$. All codewords are then on a sphere in $\mathbb{R}^n$, so you are choosing the one with the least angle to y. 'Matched filters' are physically realizable ways of evaluating $x^\top y$ for the various codewords x, and so are used to help make this choice.

Lemma 3.6.2.

$$P\big(f(Y|\gamma(u')) \geq f(Y|\gamma(u)) \mid U = u\big) = 1 - \Phi\left(\frac{|\gamma(u') - \gamma(u)|}{2\sqrt{v}}\right).$$

Proof. You have $f(y|x') \geq f(y|x)$ **iff** $|y - x|^2 \geq |y - x'|^2$, and (see Exercise 1)

$$|y - x|^2 - |y - x'|^2 = 2(x' - x)^\top (y - x) - |x' - x|^2. \tag{3}$$

Write $\delta := x' - x$, then

$$f(Y|x') \geq f(Y|x) \iff \delta^\top(Y - x) \geq \tfrac{1}{2}|\delta|^2$$

$$\iff \frac{1}{|\delta|\sqrt{v}}\delta^\top(Y - x) \geq \frac{|\delta|}{2\sqrt{v}}.$$

The point of bringing the inequality into this form is that, given $X = x$,

the Y_t are independent $N(x_t, v)$,

$$\therefore \quad \delta_t(Y_t - x_t) \text{ are independent } N(0, \delta_t^2 v),$$

$$\therefore \quad \delta^\top(Y - x) = \sum_t \delta_t(Y_t - x_t) \text{ is } N\left(0, \sum_t \delta_t^2 v\right),$$

$$\therefore \quad Z := \frac{1}{|\delta|\sqrt{v}}\delta^\top(Y - x) \text{ is } N(0, 1).$$

Thus

$$P(f(Y|x') \geq f(Y|x) \mid X = x) = P\left(Z \geq \frac{|\delta|}{2\sqrt{v}} \,\middle|\, X = x\right) = 1 - \Phi\left(\frac{|\delta|}{2\sqrt{v}}\right).$$

The claimed formula follows. $\square$

Lemma 3.6.3. *For equidistributed source-messages in $\mathcal{U}$, where $\#\mathcal{U} = m$, and for the ML decoder, the MGC has*

$$\beta(\gamma) \leq \frac{1}{m} \sum_u \sum_{\{u' : u' \neq u\}} e^{-|\gamma(u) - \gamma(u')|^2/(8v)}.$$

(As before, equidistribution is the worst case, by Lemma 3.1.1, so it will suffice to prove this bound for that case.)

Proof. By Lemmas 3.5.1 and 3.6.2,

$$\beta(\gamma) \leq \frac{1}{m} \sum_{u \neq u'} \sum \left(1 - \Phi\left(\frac{|\gamma(u) - \gamma(u')|}{2\sqrt{v}}\right)\right).$$

Apply Lemma 3.6.1 to this. □

Now code, i.e. choose m codewords from the set C_n of n-strings of real numbers. In effect, choose m points in $C_n \subseteq \mathbb{R}^n$. With unrestrained choice, obviously you can make $\beta(\gamma)$ as small as you want. In practice your choice is restricted by one or other of the following physical requirements:

(C1) Each codeword $w = w_1 w_2 \ldots w_n$ is to have $w_t^2 \leq p$ for all t (a bound on peak signal-amplitude). Here $C_n = [-\sqrt{p}, \sqrt{p}]^n$.

(C2) Each codeword $w = w_1 w_2 \ldots w_n$ is to have $(w_1^2 + \cdots + w_n^2)/n \leq p$ (a bound on signal power). Here $C_n = \{w : (w_1^2 + \cdots + w_n^2)/n \leq p\}$.

We now calculate lower bounds for C under each of these constraints. Even though explicit formulae for C will be given in the next chapter, we consider the present direct approach to establishing that the channel can transmit reliably to be worthwhile. Under constraint (C1), in addition, no simple expression for C emerges from the next chapter's results, as you will see.

Reliable transmission

Theorem 3.6.4. *Under the peak-amplitude constraint (C1) the MGC defined by (2) has capacity C satisfying*

$$C \geq \log \frac{2}{1 + e^{-p/(2v)}} > 0.$$

Proof. Use a random code Γ. As before, $\Gamma(u_j)$ is the random n-string $W_{j1} W_{j2} \ldots W_{jn}$, where now all the W_{jt} are to be independently $\pm\sqrt{p}$ with probabilities $\frac{1}{2}$. Pick $u \neq u'$, then $\Gamma(u)$ and $\Gamma(u')$ are strings $W_1 W_2 \ldots W_n$ and $W_1' W_2' \ldots W_n'$, say. Each $(W_t - W_t')^2$ is either $(\sqrt{p} - \sqrt{p})^2 = 0$ or $(\sqrt{p} + \sqrt{p})^2 = 4p$, both with probability $\frac{1}{2}$. So

$$E e^{-|\Gamma(u) - \Gamma(u')|^2/(8v)} = E \prod_1^n e^{-('V_t - W_t')^2/(8v)}$$

$$= \prod E \qquad \text{(independence)}$$

$$= \left(\tfrac{1}{2} + \tfrac{1}{2} e^{-4p/(8v)}\right)^n = \alpha^n$$

where $\alpha := \frac{1}{2}(1 + e^{-\frac{1}{2}p/v})$. Then

$$E\beta(\Gamma) \leq (m-1)\alpha^n = 2^{n(r+o(1))}\alpha^n$$

which will tend to 0 provided $r < -\log \alpha$. Finish as in the proof of Theorem 3.5.3.   □

Theorem 3.6.5. *Under the signal-power constraint (C2) the MGC defined by (2) has capacity C satisfying*

$$C \geq \tfrac{1}{2} \log \left(1 + \frac{p}{2v}\right) > 0.$$

Proof. This time choose p' satisfying $0 < p' < p$ and let all the W_{jt} be independent $N(0, p')$. The resulting random code Γ may have some codewords violating (C2), but we ignore that at first and just assume Γ is used as the code. Distinct source-words u, u' are thus encoded by random strings $\Gamma(u) = W_1 W_2 \ldots W_n$ and $\Gamma(u') = W_1' W_2' \ldots W_n'$, say. The r.v.s $W_t - W_t'$ are independent $N(0, 2p')$, so

$$Ee^{-|\Gamma(u)-\Gamma(u')|^2/(8v)} = E \prod_1^n e^{-(W_t - W_t')^2/(8v)}$$

$$= \prod E$$

$$= \left(\frac{1}{\sqrt{4\pi p'}} \int_{-\infty}^{\infty} e^{-u^2/(8v)} e^{-u^2/(4p')} \, du\right)^n$$

$$= \left(\frac{1}{\sqrt{4\pi p'}} \sqrt{\frac{2\pi}{\frac{1}{4v} + \frac{1}{2p'}}}\right)^n = \left(1 + \frac{p'}{2v}\right)^{-\frac{n}{2}}.$$

Thus

$$E\beta(\Gamma) \leq (m-1)\left(1 + \frac{p'}{2v}\right)^{-\frac{n}{2}} = 2^{n(r+o(1))}\left(1 + \frac{p'}{2v}\right)^{-\frac{n}{2}}.$$

The transmission rate r is to be such that $\delta := \frac{1}{2}\log(1 + \frac{1}{2}p'/v) - r > 0$. Since $m = 2^{n(r+o(1))}$ the previous line gives $E\beta(\Gamma) \leq 2^{-n(\delta+o(1))}$.

Let $\mathcal{U}^o$ be the random set of all source-words $u \in \mathcal{U}_n$ for which $\Gamma(u)$ violates constraint (C2). Define a modified random code Γ' by

$$\Gamma'(u) := \begin{cases} \Gamma(u), & \text{if } u \notin \mathcal{U}^o, \\ 0, & \text{if } u \in \mathcal{U}^o. \end{cases}$$

In coding a source-word u, cases when use of Γ' with the ML decoder can lead to error are covered by the possibilities

- transmitting $\Gamma(u)$ leads to error,
- $u \in \mathcal{U}^o$,
- $f(Y|0) \geq f(Y|\Gamma(u))$.

Thus cases when the random source-word U, coded by $X' := \Gamma'(U)$, may lead to error are included in the possibilities

- transmitting $X = \Gamma(U)$ leads to error,
- $(X_1^2 + \cdots + X_n^2)/n > p$, where $X = X^{(n)} = X_1 X_2 \ldots X_n$,
- $f(Y|0) \geq f(Y|X)$.

You conclude that

$$E\beta(\Gamma') \leq E\beta(\Gamma) + P\left(\frac{1}{n}\sum_1^n X_t^2 > p\right) + P(f(Y|0) \geq f(Y|X)). \quad (4)$$

Now from Lemmas 3.6.2 and 3.6.1 you have

$$P(f(Y|0) \geq f(Y|\gamma(u)) \mid U = u) = 1 - \Phi\left(\frac{|\gamma(u)|}{2\sqrt{v}}\right) \leq e^{-|\gamma(u)|^2/(8v)},$$

whence

$$P(f(Y|0) \geq f(Y|\gamma(U))) \leq \sum_u P(U = u)e^{-|\gamma(u)|^2/(8v)} = Ee^{-|\gamma(U)|^2/(8v)},$$

i.e.

$$P(f(Y|0) \geq f(Y|X)) \leq Ee^{-|X|^2/(8v)}.$$

In the random codeword $X = X^{(n)} = X_1 X_2 \ldots X_n$ the letters X_t are independent $N(0, p')$, so by a similar calculation to that for $Ee^{-|\Gamma(u)-\Gamma(u')|^2/(8v)}$ earlier you find that this bound is $\left(1 + \frac{p'}{4v}\right)^{-\frac{n}{2}}$. The third term on the right of (4) thus tends to 0 as $n \to \infty$.

The second does likewise, since the WLLN (§2.5) gives $n^{-1}\sum_1^n X_t^2 \xrightarrow{P} p' < p$. We bounded the first term by $2^{-n(\delta+o(1))}$, which also tends to 0, so finally you obtain $E\beta(\Gamma') \to 0$ as $n \to \infty$.

The random code Γ' is distributed over actual codes that satisfy (C2). By Lemma 3.3.1(i), for each n there is one of these with error probability at most $E\beta(\Gamma')$. Thus reliable transmission is possible at rate r by codes satisfying (C2). Since r can be any number less than $\frac{1}{2}\log(1 + \frac{1}{2}p'/v)$ it follows that the supremum C of the r-values is at least $\frac{1}{2}\log(1 + \frac{1}{2}p'/v)$. Then since, in turn, p' can have any value below p, you finally conclude the result. $\square$

The capacity when constraint (C2) is in force is actually $C = \frac{1}{2}\log(1 + p/v)$, plotted as the upper curve in Fig. 3.6.2. The other curve is the bound we have just calculated.

The capacity is thus a function of the *signal-to-noise ratio* p/v, since v represents the average noise-power just as p is the maximum allowed average signal-power.

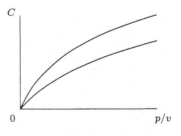

Figure 3.6.2

Exercises

1.   Verify (3).
       Hint: write $y - x'$ as the difference between $y - x$ and $x' - x$ and expand
       by the 'cosine rule for triangles'.

2.   Prove Mills's ratio (1) by establishing that
       $$(x^{-1} - x^{-3})\phi(x) < 1 - \Phi(x) < x^{-1}\phi(x) \qquad (x > 0).$$
       Hint: try differentiating.

CHANNEL CODING THEOREMS

The aim will be to evaluate channel capacity C and find how in principle it may be realized. Under regularity conditions we shall obtain the evaluation

$$C = \lim_{n \to \infty} \sup_{p_{X^{(n)}}} \frac{1}{n} I(X^{(n)} \wedge Y^{(n)}),$$

where $I(X^{(n)} \wedge Y^{(n)})$ is the *mutual information* between channel input $X^{(n)}$ and output $Y^{(n)}$. Mutual information will be defined and investigated in §4.1. Here it represents the number of bits that knowledge of $Y^{(n)}$ contributes to knowledge of $X^{(n)}$, and the $1/n$ converts it into per-symbol form. The supremum is over all probability distributions for $X^{(n)}$, and represents 'channel matching': you are taking a probability distribution best suited to the channel. Finally the limit in n is because to transmit at a rate close to capacity you may need to encode long blocks of source output as single letters.

We consider *all* distributions for X; how does that square with our considering only equidistributed U? Answer: the best p_X may well be equidistribution *over some subset of C_n suited to the channel.*

Example 4.0.1. Suppose a binary channel is twice as likely to transmit 1 incorrectly as 0. Common sense says: try to send half as many 1s through the channel as 0s. So send only words with twice as many 0s as 1s. One can have equidistribution over the class of such words, e.g. over the $\binom{6}{2} = 15$ such words of length 6.

4.1 Mutual information

Definition. *The* mutual information *between jointly distributed discrete* r.v.s X, Y is

$$I(X \wedge Y) := E \log_+ \frac{p(X,Y)}{p_X(X)p_Y(Y)} \tag{1}$$

$$= \sum_x \sum_y p(x,y) \log_+ \frac{p(x,y)}{p_X(x)p_Y(y)}$$

$$= h(X) + h(Y) - h(X,Y).$$

The symmetry of I is often considered surprising, as witness Csiszár & Körner (1981), p. 21 ("remarkable fact"), or Welsh (1988), p. 11 ("... as far as I can see, has no intuitive explanation"). There is, though, about this point as many others, an illuminating discussion in Rényi (1987), pp. 24–25, 33, a posthumously-published work by the Hungarian probabilist and information-theorist Alfred Rényi. He links the symmetry of I, tentatively, with its lack of dependence on any *causal* relationship between its arguments.

Although I is symmetric in its arguments it is often needed in situations of asymmetry, such as when X is an input to a channel and Y the corresponding output. For calculation in such cases the formulae

$$I(X \wedge Y) = E \log_+ \frac{f(Y|X)}{p_Y(Y)} = \sum_x \sum_y p_X(x)f(y|x) \log_+ \frac{f(y|x)}{p_Y(y)}$$

are useful.

From Proposition 2.10.1,

$$I(X \wedge Y) = h(Y) - h(Y|X), \tag{2}$$

so I measures the amount of uncertainty about Y that is removed by X. Or, I is the 'information about Y conveyed by X'. Because I is symmetric you can interchange X and Y in these slogans and formulae.

For the first main result on mutual information we need the notational convention that '$\wedge$' binds less tightly than ','. So you read $I(X, Y \wedge Z)$ as $I((X,Y) \wedge Z)$.

Theorem 4.1.1.

$$I(X \wedge Y) \geq 0, \text{ with equality iff } X, Y \text{ are independent.} \tag{3}$$

$$I(X, Y \wedge Z) \geq I(Y \wedge Z), \tag{4}$$

with equality iff X and Z are conditionally independent given Y.

Proof. (3) is Corollary 2.10.3. (4) follows from (2) and (2.10.7). □

Formula (3) implies that $I(X \wedge Y)$ is a good measure of the *dependence* of r.v.s X and Y, better in many respects than correlation.

To understand (4), recall from §2.9 the three characterisations of conditional independence presented there: the definition, redundant conditioning, and Markov chain. In relation to communication channels a fourth characterisation is the most important: conditional independence of X and Z given Y precisely encapsulates the situation when X is an input to a first channel, the output Y from which is used as input to a second, independently operating, channel, the final output from which is Z, as illustrated.

$$X \xrightarrow{\quad} \boxed{\text{First channel}} \xrightarrow{\quad Y\quad} \boxed{\text{Second channel}} \xrightarrow{\quad Z\quad}$$

The two channels linked in this way are said to be in *cascade*, and together to constitute a *cascade* channel. The equality case of (4) thus holds if and only if X, Y and Z can arise in this way from a cascade channel.

Other properties of cascade channels are explored in the Exercises and Problems.

You can express (4) alternatively as a 'pooling inequality'

$$I(Y \wedge Z) \geq I(\psi(Y) \wedge Z). \tag{5}$$

Finally, note from (2.10.10) that

$$I(X^{(n)} \wedge Y^{(n)}) \geq h(X^{(n)}) - \sum_{1}^{n} h(X_t | Y_t).$$

Thus, if X_1, X_2, ..., are mutually independent,

$$I(X^{(n)} \wedge Y^{(n)}) \geq \sum_{1}^{n} I(X_t \wedge Y_t). \tag{6}$$

The Hu correspondence

The mass of identities and inequalities in §2.9 and above will doubtless seem indigestible. There is however a curious way to write down correct identities and inequalities involving h and I, as follows. In expressions involving h and/or I replace random variables X, Y, ... by sets A, B, ... and replace h and I by an arbitrary non-negative additive set-function μ. Think of μ as 'area', or 'volume', or 'probability' of its set-argument. Make also the following substitutions of symbols:

$$, \leftrightarrow \cup$$
$$| \leftrightarrow \setminus$$
$$\wedge \leftrightarrow \cap$$

You get correspondences such as

$$h(X) \leftrightarrow \mu(A),$$
$$h(X,Y) \leftrightarrow \mu(A \cup B),$$
$$h(Y|X) \leftrightarrow \mu(B \setminus A),$$
$$h(Z|X,Y) \leftrightarrow \mu(C \setminus (A \cup B)),$$
$$I(X \wedge Y) \leftrightarrow \mu(A \cap B),$$
$$I(X,Y \wedge Z) \leftrightarrow \mu((A \cup B) \cap C).$$

Then *any linear identity or inequality involving h and/or I is valid iff the corresponding formula*

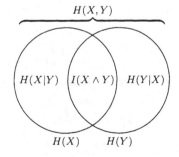

Figure 4.1.1

for μ is valid. This is due to Hu Kuo-Ting (1962); for a textbook proof see Csiszár & Körner (1981), p. 52.

Relations such as (2) and those in Proposition 2.10.1 can then be pictured in a Venn-like diagram (Fig. 4.1.1).

Exercises

1.   What is $I(X \wedge X)$?

2.   A source has a 4-letter alphabet and emits successive letters independently. The distribution of emitted letters is shown in Table 4.1.2, together with three decipherable binary codes I, II, III for the source. For each code, what is the mutual information between the source letter and the first symbol of the codeword?

Table 4.1.2

u	p_u	I	II	III
1	·4	1	0	0
2	·3	01	01	01
3	·2	001	011	011
4	·1	000	111	0111

3.   Let X have p.m.f. $(p_X(x))_{x \in \mathcal{X}}$ and let Y have conditional p.m.f. $(f(y|x))_{y \in \mathcal{Y}}$ given $X = x$. Here $\mathcal{X}$, $\mathcal{Y}$ are finite sets. Show that $I(X \wedge Y)$ is the infimum of

$$\sum_{x} p_X(x) f(y|x) \log_+ \frac{f(y|x)}{q(y)}$$

over all p.m.f.s $q(\cdot)$ on $\mathcal{Y}$.

4.   Consider a BSC (§3.4) with error probability $\frac{1}{10}$. A codeword to be sent through it is either 00 or 11, these being equally likely. Calculate the mutual information between the codeword sent and the first letter received as output form the channel. Find how much extra mutual information accrues with the receipt of the second letter output from the channel.

5.   Let X take values in $\mathcal{X}$ and Y in $\mathcal{Y}$. For a specified conditional p.m.f. $f(\cdot|x)$ for Y given X, write $I(p)$ for $I(X \wedge Y)$ when X has distribution $p = (p(x))_{x \in \mathcal{X}}$. Write $q = (q(y))_{y \in \mathcal{Y}}$ for the resulting distribution of Y, so that $q(y) = \sum_x p(x) f(y|x)$.

(a) Suppose p and p' are two distributions for X, giving rise to corresponding distributions q and q' for Y. Let $0 \le \lambda \le 1$, and form linear combinations of distributions as in Exercise 2.9.6. Show that

$$I(\lambda p + (1 - \lambda)p') - \lambda I(p) - (1 - \lambda)I(p')$$
$$= h(\lambda q + (1 - \lambda)q') - \lambda h(q) - (1 - \lambda)h(q').$$

(b) By using the result of Exercise 2.9.6 deduce that $I(p)$ is a concave function of p:
$$I(\lambda p + (1 - \lambda)p') \geq \lambda I(p) + (1 - \lambda)I(p').$$
(c) Hence show by induction that if $p^{(1)}, \ldots, p^{(m)}$ are distributions for X, and $\lambda_1, \ldots, \lambda_m$ are non-negative numbers that sum to 1, then
$$I(\lambda_1 p^{(1)} + \cdots + \lambda_m p^{(m)}) \geq \lambda_1 I(p^{(1)}) + \cdots + \lambda_m I(p^{(m)}).$$

6. Let X, Y_1 and Y_2 be discrete r.v.s and let Y be a r.v. whose conditional distribution given X is a linear combination of those of Y_1 given X, and Y_2 given X:
$$f_{Y|X}(y|x) = \lambda f_{Y_1|X}(y|x) + (1 - \lambda)f_{Y_2|X}(y|x) \qquad \text{(all } y, x),$$
where $\lambda \in [0,1]$ is a constant. Prove that
$$I(X \wedge Y) \leq \lambda I(X \wedge Y_1) + (1 - \lambda)I(X \wedge Y_2).$$
This shows that $I(U \wedge V)$ is convex as a function of the conditional distribution of V given U.
   Hint: Show that
$$\lambda I(X \wedge Y_1) + (1 - \lambda)I(X \wedge Y_2) - I(X \wedge Y)$$
$$= \lambda \sum_x \sum_y f_{X,Y_1}(x, y) \log_+ \frac{f_{X|Y_1}(x|y)}{f_{X|Y}(x|y)} +$$
$$+ (1 - \lambda) \sum_x \sum_y f_{X,Y_2}(x, y) \log_+ \frac{f_{X|Y_2}(x|y)}{f_{X|Y}(x|y)},$$
and apply the Gibbs inequality.

4.2   The Second Coding Theorem (SCT)

We take up again the notational framework for channels established in Chapter 3. In particular, $X^{(n)} = \gamma_n(U^{(n)})$ where $U^{(n)}$ is random (equiprobable) over $\mathcal{U}_n = \{u_1, \ldots, u_m\}$. Here $U^{(n)}$ is the source-word, γ_n is the code, and $X^{(n)}$ is the codeword transmitted through the channel. For certain proofs we replace γ_n by the random code Γ_n which is equidistributed over the set of all functions from $\mathcal{U}_n$ to some set $\mathcal{C}_n$. Where no confusion results we shall omit the sub- or superscript n, as we did in Chapter 3. Output from the channel is $Y = Y^{(n)}$ which depends (randomly) only on X, through the channel-specified conditional distribution $P(Y = y|X = x) = f(y|x)$. Until we get to §4.5, on 'continuous entropy', all r.v.s are discrete, and we denote the joint and 'marginal' distributions of X and Y by $p(\cdot, \cdot)$, $p_X(\cdot)$, $p_Y(\cdot)$ as in §§2.9–10.

Set

$$C_n := \sup_{P_{X^{(n)}}} \frac{1}{n} I(X^{(n)} \wedge Y^{(n)}).$$

Theorem 4.2.1: converse part of the SCT. $C \leq \limsup_{n \to \infty} C_n$.

Proof. Let $\gamma_n : \mathcal{U}_n \to \mathcal{C}_n$ be a code, where $\#\mathcal{U}_n = m = 2^{n(r+o(1))}$ for some r. We prove the following lower bound on the error probability:

$$\beta(\gamma_n) \geq 1 - \frac{C_n + \frac{1}{n}}{r + o(1)}. \tag{1}$$

If γ_n is non-injective we can alter it to make it injective, without increasing the error probability. So it suffices to prove (1) for injective γ_n. In that case $X = X^{(n)} = \gamma(U)$ is equidistributed over some set $\mathcal{X}_n$ of m elements; also it suffices to guess X, rather than U, from Y. Let our guess be $d(Y)$ where $d(\cdot)$ is a decoder. Then

$$
\begin{aligned}
nC_n &\geq I(X \wedge Y) \\
&\geq I(X \wedge d(Y)) \qquad \text{(by (4.1.5))} \\
&= h(X) - h(X|d(Y)) \\
&= \log m - h(X|d(Y)) \qquad \text{(equidistribution)} \\
&\geq \log m - 1 - \beta(\gamma_n) \log(m-1) \qquad \text{(by Theorem 2.10.4)}
\end{aligned}
$$

and (1) follows.

For r to be a reliable transmission rate, i.e. $\beta(\gamma_n) \to 0$, you must then have $r \leq \limsup C_n$. So $C \leq \limsup C_n$.   □

For the direct part of the SCT we need two lemmas.

Lemma 4.2.2. *Let Γ be a random coding (as always, independent of U) such that the codewords $\Gamma(u_1)$, ..., $\Gamma(u_m)$ are independent and each has the same distribution p_W, say (so $p_W(w) := P(\Gamma(u_1) = w)$). Define W_1, ..., W_{m-1} by:*

$$\text{if } U = u_j \text{ then } W_i := \begin{cases} \Gamma(u_i) & \text{for } i < j \text{ (if any)}, \\ \Gamma(u_{i+1}) & \text{for } i \geq j \text{ (if any)}. \end{cases}$$

Then $U, X, W_1, W_2, \ldots, W_{m-1}$ are independent, and each of X, W_1, W_2, ..., W_{m-1} has distribution p_W. (So $p_X = p_W$.)

The idea here is that U picks out one element of the list $\Gamma(u_1)$, ..., $\Gamma(u_m)$ to become $X = \Gamma(U)$, and the others are re-indexed from 1 to $m-1$ as $W_1, \ldots, W_{m-1}$:

$$
\begin{array}{ccccccc}
& & & X & & & \\
& & & \uparrow & & & \\
\Gamma(u_1) & \cdots & \Gamma(u_{j-1}) & \Gamma(u_j) & \Gamma(u_{j+1}) & \cdots & \Gamma(u_m) \\
\downarrow & & \downarrow & & \downarrow & & \downarrow \\
W_1 & \cdots & W_{j-1} & & W_j & \cdots & W_{m-1}
\end{array}
$$

Proof.

$$
P(U = u_j, X = x, W_1 = w_1, \ldots, W_{m-1} = w_{m-1})
$$

$$
= P\left(U = u_j, \begin{pmatrix} \Gamma(u_1) \\ \vdots \\ \Gamma(u_m) \end{pmatrix} = \begin{pmatrix} w_1 \\ \vdots \\ w_{j-1} \\ x \\ w_j \\ \vdots \\ w_{m-1} \end{pmatrix} \right)
$$

$$
= P(U = u_j) p_W(x) p_W(w_1) p_W(w_2) \cdots p_W(w_{m-1}).
$$

$\square$

Define the r.v.
$$
\eta_n := \frac{1}{n} \log_+ \frac{p(X,Y)}{p_X(X) p_Y(Y)} \quad \left(= \frac{1}{n} \log_+ \frac{p(X^{(n)}, Y^{(n)})}{p_{X^{(n)}}(X^{(n)}) p_{Y^{(n)}}(Y^{(n)})} \right),
$$
so that, by (4.1.1),
$$
E\eta_n = n^{-1} I(X^{(n)} \wedge Y^{(n)}).
$$

Lemma 4.2.3. *Fix n. For Γ as in the previous lemma, and any real t,*
$$
E\beta(\Gamma) \leq P(\eta_n \leq t) + m 2^{-nt}. \tag{2}
$$

Proof. Let $S_y(x) := \{x' \in C_n : f(y|x') \geq f(y|x)\}$. As we argued in the proof of Lemma 3.2.4, this set includes all n-strings that, if they are codewords, can lead to a mistake if x is sent, y is received and ML decoding is used. Consider the indicator function
$$
e(\gamma, u, y) := \mathbf{1}\{\gamma(u') \in S_y(\gamma(u)) \text{ for some } u' \neq u\}.
$$
Then
$$
e(\gamma, u, y) = 1 - \prod_{\{u' : u' \neq u\}} \mathbf{1}\{\gamma(u') \notin S_y(\gamma(u))\}
$$
$$
= 1 - \prod_{\{u' : u' \neq u\}} (1 - \mathbf{1}\{\gamma(u') \in S_y(\gamma(u))\}). \tag{3}
$$
For any *specific* coding γ,
$$
\beta(\gamma) \leq E e(\gamma, U, Y)
$$

and so
$$E\beta(\Gamma) \le Ee(\Gamma, U, Y).$$
In the right-hand side of (3) we thus need to replace γ, u, y respectively by Γ, U, Y and take expectation. In the notation of Lemma 4.2.2 we find
$$E\beta(\Gamma) \le E\left(1 - \prod_{i=1}^{m-1}(1 - \mathbf{1}\{W_i \in S_Y(X)\})\right).$$
The right-hand side equals
$$\sum_x \sum_y p_X(x)f(y|x)E\left(1 - \prod_{i=1}^{m-1}(1 - \mathbf{1}\{W_i \in S_y(x)\}) \,\Big|\, X = x, Y = y\right)$$
(in other words, $\sum_x \sum_y P(X = x, Y = y)E(\cdot|X = x, Y = y)$, similarly to (2.10.1)). But, by Lemma 4.2.2,
$$E\left(\prod_{i=1}^{m-1}(1 - \mathbf{1}\{W_i \in S_y(x)\}) \,\Big|\, X = x, Y = y\right) = \prod_{i=1}^{m-1} E(1 - \mathbf{1}\{W_i \in S_y(x)\})$$
$$= (1 - Q_y(x))^{m-1},$$
where $Q_y(x) := \sum_{x' \in S_y(x)} p_X(x')$. So, at last,
$$E\beta(\Gamma) \le 1 - (1 - Q_Y(X))^{m-1}.$$
Denote the set of (x, y) such that
$$\frac{1}{n}\log_+ \frac{p(x,y)}{p_X(x)p_Y(y)} > t$$
by T, and use the bounds
$$1 - (1 - Q_y(x))^{m-1} \le \begin{cases} mQ_y(x), & \text{in } T, \\ 1, & \text{elsewhere;} \end{cases}$$
then
$$E\beta(\Gamma) \le P((X, Y) \notin T) + m\sum\sum_{(x,y) \in T} p_X(x)f(y|x)Q_y(x). \qquad (4)$$
The first term on the right is just $P(\eta_n \le t)$.

Now for (x, y) in T and x' in $S_y(x)$ you have
$$f(y|x') \ge f(y|x) \qquad \text{(definition of } S_y(x))$$
$$\ge p_Y(y)2^{nt} \qquad \text{(definition of } T).$$
Multiply by $p_X(x')/p_Y(y)$ to get
$$P(X = x'|Y = y) \ge p_X(x')2^{nt}.$$
Summing over $x' \in S_y(x)$, you deduce that
$$1 \ge Q_y(x)2^{nt},$$
so the double sum in (4) is bounded by 2^{-nt}. $\square$

Theorem 4.2.4: direct part of the SCT. *Suppose that there exists a constant $c > 0$ such that for every r with $0 < r < c$ we can carry out the following construction: with $m = m_n = 2^{n(r+o(1))}$, for each n find a random coding $\Gamma = \Gamma_n$ with the property that the codewords $\Gamma(u_1), \ldots, \Gamma(u_m)$ are i.i.d. r.v.s, such that*

$$\eta_n \xrightarrow{P} c \qquad (n \to \infty).$$

Then $C \geq c$.

Proof. Fix $\varepsilon > 0$ and take $r := c - 2\varepsilon$ and $t := c - \varepsilon$ in (2); then

$$E\beta(\Gamma_n) \leq P(\eta_n \leq c - \varepsilon) + 2^{-n(\varepsilon+o(1))}$$
$$\to 0$$

as $n \to \infty$. So by Lemma 3.3.1(i) there is a code γ_n for each n such that the error probabilities $\beta(\gamma_n)$ converge to 0. Thus reliable transmission is possible at rate r. Since this holds for every $r < c$, the supremum C of the r-values is at least c. □

We have now established the bounds

$$c \leq C \leq \limsup C_n$$

on capacity C, and will establish identity of the two bounds in particular cases.

4.3 The discrete memoryless channel (DMC)

$X^{(n)}$ is a random n-string $X_1 X_2 \ldots X_n$ of input letters and $Y^{(n)}$ a random n-string $Y_1 Y_2 \ldots Y_n$ of output letters. The superscripts will now *not* be omitted from $X^{(n)}$ and $Y^{(n)}$.

Lemma 4.3.1. *For a DMC,*

$$I(X^{(n)} \wedge Y^{(n)}) \leq \sum_1^n I(X_t \wedge Y_t),$$

with equality if the X_t are mutually independent.

(Contrast this with (4.1.6), where the inequality is reversed, but is proved only under independence.)

Proof. $P(Y^{(n)} = y_1 y_2 \ldots y_n | X^{(n)} = x_1 x_2 \ldots x_n) = \prod_1^n p_{x_t y_t}$, hence

$$h(Y^{(n)}|X^{(n)}) = \sum_1^n h(Y_t|X_t).$$

So

$$I(X^{(n)} \wedge Y^{(n)}) = h(Y^{(n)}) - h(Y^{(n)}|X^{(n)})$$
$$= h(Y^{(n)}) - \sum_1^n h(Y_t|X_t)$$
$$\leq \sum_1^n h(Y_t) - \sum_1^n h(Y_t|X_t) = \sum_1^n I(X_t \wedge Y_t).$$

Equality holds if the Y_t are independent, which is so if the X_t are independent.   □

Theorem 4.3.2. *For a DMC the capacity is $\bar{C}$ defined by*

$$\bar{C} := \sup_{p_{X_1}} I(X_1 \wedge Y_1), \tag{1}$$

where X_1 and Y_1 are input and output letters, and the supremum is over all distributions of X_1. The codewords $W_j = W_{j1} W_{j2} \ldots W_{jn}$ of an asymptotically optimal random coding can be obtained by choosing the mn letters independently with the distribution $\bar{p}$, where $\bar{p}$ is the maximizing p_{X_1} in (1).

Proof. By Lemma 4.3.1,

$$nC_n \leq \sup_{p_{X^{(n)}}} \sum_{t=1}^n I(X_t \wedge Y_t)$$
$$\leq \sum_1^n \sup I(X_t \wedge Y_t) = n\bar{C}.$$

Since, by Theorem 4.2.1, the capacity C is bounded above by $\limsup C_n$, we conclude $C \leq \bar{C}$. But if we employ the random coding suggested then

$$\eta_n = \frac{1}{n} \sum_{t=1}^n \log_+ \frac{p_{X_t, Y_t}(X_t, Y_t)}{p_{X_t}(X_t) p_{Y_t}(Y_t)},$$

and the summands are i.i.d. with expectation $I(X_1 \wedge Y_1) = \bar{C}$. The Weak Law of Large Numbers (§2.5) says $\eta_n \xrightarrow{P} \bar{C}$, so $C \geq \bar{C}$ by Theorem 4.2.4. The capacity is thus $\bar{C}$.

To establish the asymptotic optimality of the random coding, argue as follows. Let Γ_n be the random coding applied to n-blocks. The second line of proof of Theorem 4.2.4 says $\bar{\beta}_n \to 0$ where $\bar{\beta}_n := E\beta(\Gamma_n)$. By Lemma 3.3.1(ii),

$$P\left(\beta(\Gamma_n) \leq \frac{\bar{\beta}_n}{1-\rho}\right) \geq \rho \qquad (0 \leq \rho \leq 1).$$

Take $\rho := 1 - \sqrt{\bar{\beta}_n}$, then

$$P\left(\beta(\Gamma_n) \le \sqrt{\bar{\beta}_n}\right) \ge 1 - \sqrt{\bar{\beta}_n} \to 1 \qquad (n \to \infty).$$

Given $\varepsilon > 0$, you can thus find $n_0 = n_0(\varepsilon)$ such that $\sqrt{\bar{\beta}_n} < \varepsilon$ for all $n \ge n_0$. For each such n, among the codes γ_n that Γ_n can be there is a set A_n of them, with $P(\Gamma_n \in A_n) > 1 - \varepsilon$, such that $\beta(\gamma_n) < \varepsilon$ for every $\gamma_n \in A_n$. $\square$

The reason for our introducing the random coding is that its symmetric definition over all codes allows us to bound the average error-probability sufficiently accurately (Lemma 4.2.3). Then Lemma 3.3.1(i) gives existence of at least one non-random code $\gamma = \gamma_n$ for each n with error rate at least as good as the average. Indeed, more than that, *nearly all* codes randomly generated using $\bar{p}$ as above achieve capacity in the limit $n \to \infty$, as the last part of the above proof establishes.

Exercises

1. *DMCs in cascade.* Cascades were introduced in §4.1. Two DMCs are in cascade if the output alphabet of the first is included in the input alphabet of the second, and the output of the first channel is fed letter-by-letter into the second, with no processing. Find the channel matrix of the resulting 'cascade' DMC in terms of the channel matrices of its component DMCs.

2. Show that the capacity of a DMC is positive unless all rows of the channel matrix are identical.

   By Exercise 3.4.3(a) this is equivalent to the channel being useless as defined in §3.4.

3. Exhibit a non-useless DMC for which there are two different input-distributions p and p' for X_t that are optimal in that they attain the supremum in (1). Then apply Exercise 4.1.5 to deduce that $\lambda p + (1 - \lambda)p'$ is an optimal input-distribution for every λ satisfying $0 \le \lambda \le 1$.

4. Suppose q and q' are two distributions for the output letter Y from a DMC, and that for each of them a corresponding input-distribution attains the supremum in (1). Use Exercise 4.1.5(a) to show that $h(\lambda q + (1 - \lambda)q') = \lambda h(q) + (1 - \lambda)h(q')$ for all $0 \le \lambda \le 1$. Then apply Exercise 2.9.7 to deduce that $q = q'$. Thus *a DMC has a unique optimal distribution for output letters.*

5.   Show that the capacity of the channel with channel matrix
$$\begin{pmatrix} \frac{2}{3} & \frac{1}{3} & 0 \\ \frac{1}{3} & \frac{1}{3} & \frac{1}{3} \\ 0 & \frac{1}{3} & \frac{2}{3} \end{pmatrix}$$
may be realized by taking the probability of one of the input letters
to be 0. Explain why that is so, and calculate the channel capacity.
       Hint: $I(X \wedge Y) \le h(Y) \le \log 3$, and there is an input distribution for
which $I(X \wedge Y)$ attains the value $\log 3$.

6.   Consider the DMC shown in Fig. 4.3.1.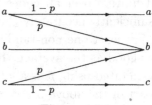
Show that if $p \le \frac{1}{3}$ then there is a dis-
tribution for the input letter X such
that the corresponding output-letter Y
is equidistributed over $\{a, b, c\}$. Deduce that for such p the channel's capacity is $\log 3$.

Figure 4.3.1

       Can you find the capacity when $\frac{1}{3} < p \le 1$?

7.   Suppose you have two independent DMCs, with capacities C_1, C_2
bits/sec. Consider the capacity C of a compound channel formed as
stated in the following three ways.
       (a) *The product channel.* Let the channels be used in parallel, in the
sense that at every second a symbol is transmitted through channel
1 (from its input alphabet) and a symbol through channel 2 (from its
input alphabet); each channel thus emits one symbol each second.
Show that $C = C_1 + C_2$.
       (b) If the channels have the same input alphabet and at each second
a symbol is chosen and sent simultaneously down both channels,
show that $C \ge \max(C_1, C_2)$, and give an example to show that
equality need not occur.
       (c) *The sum channel.* If channel i has matrix Π_i and the compound
channel has
$$\Pi = \begin{pmatrix} \Pi_1 & 0 \\ 0 & \Pi_2 \end{pmatrix},$$
show that C is given by $2^C = 2^{C_1} + 2^{C_2}$. To what mode of operation
does this correspond?

4.4   Symmetric channels

Recall from §3.4 that a symmetric channel is one whose channel matrix has all rows permutations of each other, and all columns permutations of each other. As before, denote the codewords of the random coding by $W_j = W_{j1}W_{j2}\ldots W_{jn}$, for $j = 1, \ldots, m$.

Theorem 4.4.1. *A symmetric channel with l-symbol output alphabet has capacity*

$$C = \log l + \sum_{j=1}^{l} p_j \log p_j$$

where $(p_j)^{\top}$ is the first row of the channel matrix. The capacity is realized by a random coding with independent W_{jt}, each equidistributed over the input alphabet.

Proof. For each j, $\sum_k p_{jk} \log p_{jk} = \sum_k p_k \log p_k$ on rearranging. By (2.10.2), $h(Y_1|X_1) = -\sum_j P(X = j) \sum_k p_{jk} \log p_{jk}$, which thus equals $-\sum_k p_k \log p_k$. Then

$$I(X_1 \wedge Y_1) = h(Y_1) - h(Y_1|X_1)$$
$$\leq \log l + \sum_j p_j \log p_j,$$

and the bound is realized if Y_1 is equidistributed. This happens when X_1 is equidistributed over the input alphabet, $\{1,\ldots,a\}$ say, because then

$$P(Y_1 = k) = \sum_{j=1}^{a} P(X = j)p_{jk} = \frac{1}{a}\sum_{j=1}^{a} p_{jk}$$
$$= \frac{1}{a}\sum_{j=1}^{a} q_j,$$

where (q_j) is the first column of the channel matrix. So all $P(Y_1 = k)$ are the same, that is, equidistribution. $\qquad\square$

In §3.4 we defined the BSC (binary symmetric channel) to be the symmetric channel with channel matrix $\begin{pmatrix} q & p \\ p & q \end{pmatrix}$, where $q = 1 - p$. The specialization of Theorem 4.4.1 to this case is as follows.

Theorem 4.4.2. *The BSC has capacity*

$$C = 1 + p\log p + q\log q \qquad \text{bits/unit time,}$$

achieved by a random coding in which the W_{jt} are independent, taking values 0, 1 with probabilities $\frac{1}{2}$.

Exercises

1. (a) Consider a BSC that can transmit 1200 bits per second and has

error probability 0·02. How many bits per second can it transmit reliably?

(b) Suppose a BSC with a physical capacity of 1000 bits per second can transmit 960 bits per second with arbitrarily low error-probability given suitable coding of blocks. How low, roughly, does this imply its error probability must be?

2.   Give an example of a channel matrix for each of the following sorts of channel, and calculate the capacity in each case:
   (a) lossless and deterministic ('noiseless');
   (b) symmetric and lossless but not deterministic;
   (c) lossless but neither deterministic nor symmetric;
   (d) deterministic but neither lossless nor symmetric.

3.   Consider a cascade of k binary symmetric channels, each with error probability p. The output from each BSC before the last is fed letter-by-letter into the next BSC, without processing. Show that the whole cascade is equivalent to a single BSC with error probability $\frac{1}{2}(1 - (1 - 2p)^k)$. Check that this tends to 0 as $k \to \infty$, unless p is 0 or 1.

4.5   Continuous entropy and mutual information

Continuous entropy

There is no straightforward extension of entropy to continuous r.v.s (by which we mean those having probability densities).

Consider a random variable X with a density f that is bounded and continuous, and vanishes outside some finite interval $[-A, A]$. Break $[-A, A]$ into subintervals of length Δ. For each j it is possible to choose a point x_j in the jth subinterval such that $\Delta f(x_j)$ equals the integral of f over the subinterval. Let X_Δ be a random variable that takes values x_j with respective probabilities $\Delta f(x_j)$. The distribution of X_Δ is a good approximation to that of X, in the sense that, for any continuous g, $Eg(X_\Delta) \to Eg(X)$ as $\Delta \to 0$ (Riemann integration!). But X_Δ has entropy

$$h(X_\Delta) = -\sum_j \Delta f(x_j) \log_+ \Delta f(x_j)$$

$$= -\log \Delta - \sum_j \Delta f(x_j) \log_+ f(x_j)$$

$$\to +\infty - \int_{-\infty}^{\infty} f(x) \log_+ f(x)\, dx = +\infty$$

as $\Delta \to 0$. So

$$h_c(X) := -\int_{-\infty}^{\infty} f(x) \log_+ f(x)\, dx = -E \log_+ f(X)$$

is **not** the limit of the entropies of the approximating discrete distributions. $h_c(X)$, defined as above for any r.v. for which the integral exists, is known as the *continuous entropy*, or *differential entropy*. It can take negative as well as positive values (Exercise 1), unlike entropy.

Continuous entropy does however have many properties in common with entropy. First among these is the following extension of the Gibbs inequality (Lemma 1.4.1).

Lemma 4.5.1. *Let f and g be probability densities on $\mathbb{R}$. If either $\int_{-\infty}^{\infty} f(x) \log_+ f(x)\, dx$ or $\int_{-\infty}^{\infty} f(x) \log_+ g(x)\, dx$ is finite then*

$$-\int_{-\infty}^{\infty} f(x) \log_+ f(x)\, dx \le -\int_{-\infty}^{\infty} f(x) \log_+ g(x)\, dx,$$

with equality iff $f = g$ almost everywhere (i.e. $\int_{\{x:f(x)\neq g(x)\}} dx = 0$).

Proof. The argument used for Lemma 1.4.1 can be repeated with integrals in x over the real line replacing sums in i, $f(x)$ replacing p_i, $g(x)$ replacing q_i, and the set $\mathcal{I}$ now defined as $\{x : f(x) > 0\}$. It proves that

$$\int_{-\infty}^{\infty} f(x) \log_+ \frac{g(x)}{f(x)}\, dx \le 0,$$

with equality if and only if

$$\int_{\mathcal{I}} g(x)\, dx = 1$$

and

$$\int_{\mathcal{I}} \mathbf{1}\{g(x)/f(x) \neq 1\} = 0.$$

The first half of this equality condition is equivalent to $\int_{\{x:g(x)>0\}\backslash\mathcal{I}} dx = 0$, and hence, writing $\mathcal{D}$ for the set $\{x : f(x) \neq g(x)\}$, to

$$\int_{\mathcal{D}\backslash\mathcal{I}} dx = 0. \tag{1}$$

The second half is equivalent to

$$\int_{\mathcal{D}\cap\mathcal{I}} dx = 0. \tag{2}$$

Since the left-hand sides of (1) and (2) are non-negative, (1) and (2) are together equivalent to the sum of their left-hand sides being zero, and thus to $\int_{\mathcal{D}} dx = 0$.

Now the expression

$$\int_{-\infty}^{\infty} f \log_+ g - \int_{-\infty}^{\infty} f \log_+ f \tag{3}$$

makes sense, because we assume at least one of its terms is finite, and it equals

$$\int_{-\infty}^{\infty} f \log_+ \frac{g}{f}.$$

We conclude that (3) is non-positive, and is zero iff $\int_D dx = 0$, as claimed.☐

We have introduced the phrase 'almost everywhere' above. A property holds *almost everywhere* in $\mathbb{R}$ if and only if the set S where it fails to hold is *null*: $\int_S dx = 0$. Thus S is in quite a strong sense 'small'. It could of course be empty. Equivalently, a set is null iff for every $\varepsilon > 0$ it can be contained in a finite union of intervals of total length ε. Null sets, from the theory of Lebesgue measure and integration, make precise the notion of having 'zero length'. All countable sets, for example the set of rationals, are null, as are plenty of uncountable sets such as the Cantor ternary set.

The next theorem is of value outside the use we shall make of it, as it shows the normal (Gaussian) distributions of mean 0 to be the continuous distributions on $\mathbb{R}$ of *maximum entropy*, subject to a second-moment bound.

Analogous roles for other classes of distributions are played by the exponential and geometric families (Problems 4.8.19 and 2.14.18), and there exists a general result related to Problem 2.14.19: see, for example, McEliece (1984), Problem 1.35, or Guiaşu & Shenitzer (1985).

Theorem 4.5.2. *A continuous random variable Y on $\mathbb{R}$, constrained by*

$$E(Y^2) \leq p, \tag{4}$$

where $p > 0$ is constant, has

$$h_c(Y) \leq \tfrac{1}{2} \log(2\pi e p),$$

with equality iff Y is $N(0,p)$.

Proof. Let g denote the density of Y, and let X have the $N(0,p)$ density f. Then

$$-\int_{-\infty}^{\infty} g \log_+ f = \tfrac{1}{2} \log(2\pi p) + \frac{\log e}{2p} \int_{-\infty}^{\infty} y^2 g(y)\, dy$$

which is finite, so

$$h_c(Y) = -\int_{-\infty}^{\infty} g \log_+ g$$

$$\leq -\int_{-\infty}^{\infty} g \log_+ f$$

(by the lemma, with equality iff Y is $N(0,p)$)

$$= \tfrac{1}{2}\log(2\pi p) + \frac{\log e}{2p}E(Y^2)$$

$$\leq \tfrac{1}{2}\log(2\pi ep). \qquad \square$$

Mutual information

Mutual information carries over to continuous random variables more satisfactorily. For r.v.s X, Y with (marginal) density f_Y for Y, and conditional density $f_{Y|X}(y|x)$ for Y given $X = x$, define

$$I(X \wedge Y) := E\log_+ \frac{f_{Y|X}(Y|X)}{f_Y(Y)} \tag{5}$$

when the expectation is finite. This may be written

$$I(X \wedge Y) = E\log_+ \frac{f(X,Y)}{f_X(X)f_Y(Y)}$$

where $f(\cdot,\cdot)$ is the joint density. So I retains the symmetry in its arguments,

$$I(X \wedge Y) = I(Y \wedge X) \tag{6}$$

that we noted for discrete r.v.s in §4.1.

The definition (5) is in fact appropriate for continuous or discrete X, provided Y has a density. Note that

$$I(X \wedge Y) = h_c(Y) - h_c(Y|X), \tag{7}$$

where

$$h_c(Y|X) := -E\log_+ f_{Y|X}(Y|X).$$

With a proper interpretation we can even retain the symmetry of I when X is discrete and Y continuous, but we first need to define the right-hand side of (6). Now there is for such r.v.s a mixed 'p.m.f.-density' $f(\cdot,\cdot)$ defined by

$$f(x,y) := f_{Y|X}(y|x)p_X(x),$$

where p_X is the p.m.f. of X. In terms of it we may write $I(X \wedge Y)$ as

$$I(X \wedge Y) = E\log_+ \frac{f(X,Y)}{p_X(X)f_Y(Y)}. \tag{8}$$

Now

$$P(X = x, Y \in B) = \int_B f(x,y)\,dy$$

identically. So

$$P(X = x|y - \varepsilon < Y < y + \varepsilon) = \frac{\int_{y-\varepsilon}^{y+\varepsilon} f(x,y)\,dy}{\int_{y-\varepsilon}^{y+\varepsilon} f_Y(y)\,dy}. \tag{9}$$

We assume that, except in some finite set D_Y, f is continuous in y for each

x. Then f_Y inherits this property and for every $y \notin D_Y$ the right-hand side of (9) converges to $f(x,y)/f_Y(y)$ as $\varepsilon \downarrow 0$. So it is natural to define the conditional p.m.f. $f_{X|Y}(\cdot|y)$ for X given $Y = y$ by

$$f_{X|Y}(x|y) := \frac{f(x,y)}{f_Y(y)}.$$

Further, let us set

$$h(X|Y) := -E \log_+ f_{X|Y}(X|Y)$$

and

$$I(Y \wedge X) := h(X) - h(X|Y).$$

On substituting the previous formulae into this you find that the right-hand side equals the right-hand side of (8). Symmetry of I is thus established, provided the entropies/continuous entropies of the components are finite.

Most of the inequalities of §§2.9–10 and §4.1 remain true in the present more general setting. For instance, with a view to what will be needed later, suppose Y and Z have a joint density and ψ is a function on $\mathbb{R}$ taking finitely many values. Then the pooling inequality (4.1.5),

$$I(Y \wedge Z) \geq I(\psi(Y) \wedge Z) \tag{10}$$

holds, as may be seen as follows. Let $T = \psi(Y)$, then Lemma 4.5.1 gives

$$-\int f_{Z|Y}(z|y) \log_+ f_{Z|Y}(z|y)\, dz \leq -\int f_{Z|Y}(z|y) \log_+ f_{Z|T}(z|\psi(y))\, dz.$$

Multiply by $f_Y(y)$ and integrate with respect to y, obtaining

$$-E \log_+ f_{Z|Y}(Z|Y) \leq -E \log_+ f_{Z|T}(Z|\psi(Y)).$$

But this is

$$h_c(Z|Y) \leq h_c(Z|T) = h_c(Z|\psi(Y)).$$

Subtract both sides from $h_c(Z)$ to obtain (10).

Exercises

1.   By considering the uniform density on the interval $(0,a)$, show that h_c can take every real value.

2.   Show that a continuous r.v. with finite variance cannot have continuous entropy $+\infty$, and that a r.v. with bounded density cannot have continuous entropy $-\infty$.

   Find examples of continuous r.v.s with
   (a) continuous entropy $-\infty$,
   (b) continuous entropy $+\infty$,

(c) continuous entropy undefined, i.e. $\int_{-\infty}^{\infty} f \max(\log f, 0) = \infty$ and $\int_{-\infty}^{\infty} f \min(\log f, 0) = -\infty$.

3. Deduce from Theorem 4.5.2 the following variant: a continuous r.v. Y on $\mathbb{R}$, constrained by $E((Y-a)^2) \leq b$, has $h_c(Y) \leq \frac{1}{2}\log(2\pi e b)$, with equality iff Y is $N(a, b)$-distributed.

4. Continuous entropy can be defined for an $\mathbb{R}^d$-valued random vector X with density f in exactly the same way as in $\mathbb{R}$:

$$h_c(X) := -\int_{\mathbb{R}^d} f(x) \log_+ f(x)\, dx = -E \log_+ f(X).$$

The extension of the Gibbs inequality, Lemma 4.5.1, also carries over to $\mathbb{R}^d$, with the same proof.

Use these facts to derive, by the same method as Theorem 4.5.2 or otherwise, a d-dimensional version of it: a continuous random vector $Y = (Y_1, \ldots, Y_d)^\top$ on $\mathbb{R}^d$, constrained by $E(Y_i^2) \leq p_i$ for $i = 1, \ldots d$, has

$$h_c(Y) \leq \sum_1^d \frac{1}{2}\log(2\pi e p_i),$$

with equality iff the Y_i are independent, having distributions $N(0, p_i)$ respectively.

(This permits dependence between the components of Y.)

4.6 The memoryless channel with additive white noise

Recall that we encode the message as an n-string $X^{(n)} = X_1 X_2 \ldots X_n$ and after transmission through the channel the n-string $Y^{(n)} = Y_1 Y_2 \ldots Y_n$ is received. The *general memoryless discrete-time channel with additive 'white' noise* is that in which

$$Y_t = c(X_t) + \varepsilon_t \qquad (t = 1, 2, \ldots),$$

where c is a fixed non-random function and the ε_t are i.i.d. continuous r.v.s, independent of $(X_1, X_2, \ldots)$.

Lemma 4.6.1. *For a channel of the above sort, for each t,*

$$h_c(Y_t | X_t) = h_c(\varepsilon_t),$$

so that, by (4.5.7),

$$I(X_t \wedge Y_t) = h_c(Y_t) - h_c(\varepsilon_t)$$

(*mutual information = output entropy − noise entropy*).

Note. With h replacing h_c, the same obviously holds for the similar discrete-state channel.

Proof. Let f_ε be the density of ε_1. Then the conditional density of Y_t given $X_t = x$ is

$$f(y|x) = f_\varepsilon(y - c(x)), \tag{1}$$

and

$$\begin{aligned}
h_c(Y_t|X_t) &= -E\log_+ f(Y_t|X_t) \\
&= -E\log_+ f_\varepsilon(Y_t - c(X_t)) \\
&= -E\log_+ f_\varepsilon(\varepsilon_t) = h_c(\varepsilon_t). \quad \square
\end{aligned}$$

To appreciate the relevance of the above lemma, recall that for maximum rate of transmission we want to maximize the mutual information $I(X_t \wedge Y_t)$, as the following theorem confirms. The channel determines the noise entropy $h_c(\varepsilon_t)$, so Lemma 4.6.1 allows us to conclude that we want to maximize the output entropy $h_c(Y_t)$.

In coding for this channel the set $\mathcal{C}_n$ of legal codewords may be a Cartesian product $\mathcal{C}^n$, as with constraint (C1) of §3.6 which is a 'letter-wise' constraint, or may be not of that form, as with (C2) which is a constraint on whole codewords.

Under letter-wise restrictions we have the following general result. It will enable us to deal with the 'word-wise' constraint (C2) by a further ad-hoc modification.

Theorem 4.6.2. *The above channel, with $\mathcal{C}_n = \mathcal{C}^n$ for some $\mathcal{C}$, has capacity*

$$C = \sup I(X_1 \wedge Y_1), \tag{2}$$

the supremum being over all distributions of X_1 on $\mathcal{C}$. The codewords $W_j = W_{j1}W_{j2}\ldots W_{jn}$ of an asymptotically optimal random coding can be obtained by choosing the mn letters independently, each with the maximizing distribution in (2).

Proof. A virtual word-for-word copy of the proof for the discrete case (Theorem 4.3.2 and results leading to it) is all that is needed, with minor alterations to take account of some of the r.v.s now being continuous. First, in the proof of Theorem 4.2.1 we worked entirely with a fixed non-random code $\gamma_n : \mathcal{U}_n \to \mathcal{C}_n$. In the new situation $\mathcal{U}_n$ is still a finite set with $m = 2^{n(r+o(1))}$ elements, although $\mathcal{C}_n$ will now be larger. An injective code γ_n will have image set $\gamma_n(\mathcal{U}_n) = \{\gamma_n(u) : u \in \mathcal{U}_n\}$ of m elements, and $X^{(n)} = \gamma_n(U^{(n)})$ is a discrete r.v. equidistributed over this finite set. Further, $d(Y) = d(Y^{(n)})$ takes values in this set so is a discrete r.v. even though $Y = Y^{(n)}$ itself is continuous. We re-proved (4.1.5) for these r.v.s as (4.5.10), so the proof goes through and we have the converse part of the SCT for our present case.

Lemma 4.2.2 and subsequent results bring in a random coding Γ, and, for each $u \in \mathcal{U}_n$, $\Gamma(u)$ will typically have a continuous distribution over $\mathcal{C}_n$, so $X = \Gamma(U)$ now becomes a continuous r.v. However Lemma 4.2.2 remains in force, where now each codeword $\Gamma(u_1)$, ..., $\Gamma(u_m)$ has the same distribution, with density $f_W(\cdot)$, say, which is also the density of X. The calculation in the proof becomes

$$P(U = u_j, X \leq x, W_1 \leq w_1, \ldots, W_{m-1} \leq w_{m-1})$$
$$= P(U = u_j, \Gamma(u_1) \leq w_1, \ldots, \Gamma(u_{j-1}) \leq w_{j-1},$$
$$\Gamma(u_{j+1}) \leq w_j, \ldots, \Gamma(u_m) \leq w_{m-1})$$
$$= P(U = u_j)P(\Gamma(u_1) \leq w_1) \cdots P(\Gamma(u_{j-1}) \leq w_{j-1}) \times$$
$$\times P(\Gamma(u_j) \leq x)P(\Gamma(u_{j+1}) \leq w_j) \cdots P(\Gamma(u_m) \leq w_{m-1})$$
$$= P(U = u_j) \int_{-\infty}^{w_1} f_W \cdots \int_{-\infty}^{w_{j-1}} f_W \int_{-\infty}^{x} f_W \int_{-\infty}^{w_j} f_W \cdots \int_{-\infty}^{w_{m-1}} f_W.$$

Hence $U, X, W_1, \ldots, W_{m-1}$ are independent, and each of $X, W_1, \ldots, W_{m-1}$ has density $f_W(\cdot)$.

For Lemma 4.2.3 we redefine η_n by replacing the joint and marginal probability mass functions $p(\cdot, \cdot)$, $p_X(\cdot)$, $p_Y(\cdot)$ by joint and marginal densities $f(\cdot, \cdot)$, $f_X(\cdot)$, $f_Y(\cdot)$, and the proof goes through with these substitutions and that of integrals for sums. Next, Lemma 4.3.1 is easily extended to the present setting: the conditional density $f_n(y^{(n)}|x^{(n)})$ of $Y^{(n)}$ given $X^{(n)} = x^{(n)}$ is the product $\prod_1^n f(y_t|x_t)$ where $f(\cdot|\cdot)$ is as in (1).

The final step is to modify straightforwardly the proofs of Theorems 4.2.4 and 4.3.2. We now have

$$\eta_n = \frac{1}{n} \sum_{t=1}^{n} \log_+ \frac{f(X_t, Y_t)}{f_X(X_t)f_Y(Y_t)},$$

and it remains the case that the summands are i.i.d. with expectation $I(X_1 \wedge Y_1) = \bar{C}$. Use of the Weak Law of Large Numbers, as before, completes the whole proof. $\square$

The memoryless Gaussian channel (MGC)
We considered this channel earlier, in §3.6. However we now allow a very little more generality than in the previous model (3.6.2), by permitting the

channel to scale the input by a constant factor b:

$$Y_t = bX_t + \varepsilon_t \qquad (t = 1, 2, \ldots), \tag{3}$$

where the ε_t are independent $N(0, v)$ (*Gaussian white noise*), independent of $X_1, X_2, \ldots$.

First consider the peak-amplitude constraint (C1) from §3.6, that the code-letters w_t satisfy $w_t^2 \leq p$. You see from Theorem 4.6.2 that the capacity is

$$C = \sup I(X \wedge (bX + \varepsilon)),$$

where ε is $N(0, v)$-distributed, independent of X, and the supremum is over all distributions for the r.v. X concentrated on the interval $[-\sqrt{p}, \sqrt{p}]$. No closed form for C emerges from this constrained-maximization problem, though approximations and bounds are readily derivable.

The discussion about it in §26 of Shannon & Weaver (1949) is still worth reading.

Capacity under signal-power constraint
Under the other constraint (C2) from §3.6, that the codewords $w = w_1 w_2 \ldots w_n$ satisfy $(w_1^2 + \cdots + w_n^2)/n \leq p$, we can evaluate the capacity explicitly, though a little more work is required. First we establish a 'converse' coding result, an upper bound on the capacity.

Lemma 4.6.3. *Under constraint (C2) the capacity is at most*

$$C(p) := \tfrac{1}{2} \log(1 + b^2 p/v).$$

Proof. As in the converse part of the SCT (Theorem 4.2.1), let $\gamma_n : \mathcal{U}_n \to \mathcal{C}_n$ be a code, where $\#\mathcal{U}_n = m = 2^{n(r + o(1))}$ for some r. All the codewords $\gamma_n(u_i) = w_{i1} w_{i2} \ldots w_{in}$, for $i = 1, \ldots, m$, are in $\mathcal{C}_n$, that is, have signal power $(w_{i1}^2 + \cdots + w_{in}^2)/n \leq p$. For each $t = 1, \ldots, n$, let X_t be a r.v. taking values $w_{1t}, \ldots, w_{mt}$ with equal probability, a 'random t^{th} code-letter' as one might say. Let Y_t be the output from the channel when X_t is transmitted.

Now suppose $\tilde{X}$ is a r.v. that is equally likely to be any of the X_t, and let $\tilde{Y} = b\tilde{X} + \tilde{\varepsilon}$ be the output when $\tilde{X}$ is transmitted. As $\tilde{X}$ is equally likely to be any of the mn code-letters w_{it}, you have

$$E(\tilde{X}^2) = \frac{1}{mn} \sum_{i=1}^{m} \sum_{t=1}^{n} w_{it}^2 \leq \frac{1}{m} \sum_{i=1}^{m} p = p,$$

and so, since $\tilde{\varepsilon}$ is $N(0, v)$-distributed independently of $\tilde{X}$,

$$E(\tilde{Y}^2) = E((b\tilde{X} + \tilde{\varepsilon})^2)$$
$$= b^2 E(\tilde{X}^2) + 0 + v \leq b^2 p + v.$$

Then

$$I(\tilde{X} \wedge \tilde{Y}) = h_c(\tilde{Y}) - h_c(\tilde{\varepsilon})$$

(by Lemma 4.6.1)

$$\leq \tfrac{1}{2}\log(2\pi e(b^2 p + v)) - \tfrac{1}{2}\log(2\pi e v)$$

(by Theorem 4.5.2 applied to each term)

$$= C(p).$$

The distribution of $\tilde{X}$ is a 'convex combination' of those of the X_t:

$$P(\tilde{X} = x) = \frac{1}{n}\sum_{t=1}^{n} P(X_t = x).$$

From this it follows via Exercise 4.1.5(c) that

$$I(\tilde{X} \wedge \tilde{Y}) \geq \frac{1}{n}\sum_{t=1}^{n} I(X_t \wedge Y_t).$$

On combining this with the above bound you find that

$$\frac{1}{n}\sum_{t=1}^{n} I(X_t \wedge Y_t) \leq C(p).$$

But now Lemma 4.3.1 extends readily (Exercise 3) to the present continuous case, whence

$$I(X^{(n)} \wedge Y^{(n)}) \leq nC(p).$$

From this we can deduce

$$\beta(\gamma_n) \geq 1 - \frac{C(p) + \frac{1}{n}}{r + o(1)}$$

by exactly the calculation used in the proof of Theorem 4.2.1, the formula (4.1.5) needed for the first step having been re-proved for present circumstances as (4.5.10). Finally, for r to be a reliable transmission rate you need $\beta(\gamma_n) \to 0$, and thus r can be at most $C(p)$. The capacity itself is thus at most $C(p)$. $\qquad\square$

Theorem 4.6.4. *Consider the MGC given by (3). Under the signal-power constraint, that every codeword $w = w_1 w_2 \ldots w_n$ should satisfy*

$$(w_1^2 + \cdots + w_n^2)/n \leq p,$$

where $p > 0$ is constant, it has capacity

$$C = \tfrac{1}{2}\log(1 + b^2 p/v) \quad \text{bits/source-letter.}$$

Proof. First choose p' and r satisfying $0 < p' < p$ and $0 < r < C(p')$. For each n we employ a random code $\Gamma = \Gamma_n$ whose code letters W_{jt} are independent $N(0, p')$ r.v.s. The codewords may violate (C2) but we shall be able to modify Γ later to make them conform. Under Γ, $X = \Gamma(U)$ is an n-string $X = X^{(n)} = X_1 X_2 \ldots X_n$ in which $X_1, \ldots, X_n$ are independent $N(0, p')$. So

$$E(Y_t^2) = b^2 E(X_t^2) + 0 + v = b^2 p' + v,$$

and in fact the Y_t are independent $N(0, b^2 p' + v)$. Then

$$\begin{aligned} I(X_t \wedge Y_t) &= h_c(Y_1) - h_c(\varepsilon_1) \\ &= \tfrac{1}{2} \log(2\pi e(b^2 p' + v)) - \tfrac{1}{2} \log(2\pi e v) \\ &= \tfrac{1}{2} \log(1 + b^2 p'/v) = C(p'). \end{aligned}$$

Since the transmission rate r was chosen less than $C(p')$ we know from the proof of Theorem 4.6.2 that $E\beta(\Gamma_n) \to 0$ as $n \to \infty$. We may then modify Γ_n, exactly as in the proof of Theorem 3.6.5, to a code Γ_n' that has the same rate but conforms to the constraint (C2). We deduce that a non-random code exists giving reliable transmission at rate r, again by following the proof of Theorem 3.6.5. Since r can take any value below $C(p')$, the capacity is at least $C(p')$. Since p' itself can take any value below p, the capacity is at least $C(p)$. From Lemma 4.6.3 the capacity is at most $C(p)$, so it thus is $C(p)$. $\square$

The capacity as a function of p/v was graphed as the upper curve in Fig. 3.6.2. The formula for the capacity was discovered independently by Claude Shannon and Norbert Wiener in 1948.

Exercises

1.   Suppose that the input to a noisy channel is a sequence of symbols distributed as a random variable B, where $P(B = -1) = q$, $P(B = 1) = p$, and $p + q = 1$. The output from the channel is $Y = Bx + \varepsilon$ where ε is $N(0, v)$-distributed independently of B. Here x, v, q and p are known positive constants. A decision rule $d : \mathbb{R} \to \{0, 1\}$ is wanted. That is, if y is the value of Y received, we decide that $b = d(y)$ is the value of B that was sent. The losses are 0 for a correct decision, c if we decide $b = -1$ when in fact 1 was sent, and c' if we decide $b = 1$ when -1 was sent. Find the decision rule that minimizes expected loss.

2.   A memoryless channel has input letter X, equally likely to be $+1$ or -1. The corresponding output is $Y = X + \varepsilon$ where ε has probability density $\tfrac{1}{4}$ over the interval $(-2, 2)$, and 0 elsewhere.
    (a) Find and sketch the output probability density.

(b) Calculate $I(X \wedge Y)$.

(c) Suppose the output is discretized to form U, where $U = -1$ if $Y \leq -1$, $U = 0$ if $-1 < Y < 1$, and $U = 1$ if $Y \geq 1$. Verify that $I(X \wedge U) = I(X \wedge Y)$ and interpret this finding.

3. Verify that Corollary 2.10.3 extends to continuous entropy: for r.v.s X, Y with $h_c(X)$, $h_c(Y)$ *finite*,

$$h_c(X, Y) \leq h_c(X) + h_c(Y).$$

Deduce that Lemma 4.3.1 extends to the channels treated in this section.

4. Consider a memoryless channel with additive, not necessarily Gaussian, white noise:

$$Y_t = bX_t + \varepsilon_t,$$

where $E(X_t^2) \leq p$, $E\varepsilon_t = 0$, $E(\varepsilon_t^2) \leq v$, and $h_c(\varepsilon_t) = h$, say. Show that the capacity is at most

$$\tfrac{1}{2} \log \left(\frac{b^2 p + v}{v_e} \right)$$

where $v_e := (2\pi e)^{-1} 2^{2h}$.

4.7   Further topics

Evaluation of channel capacity

Even for a DMC, finding the input distribution that maximizes $I(X \wedge Y)$ and so realizes the channel's capacity can be difficult. However the concavity of $I(X \wedge Y)$ as a function of the distribution of X (Exercise 4.1.5) is helpful, as it allows the *Kuhn-Tucker theory of constrained optimization* to come into play. That theory establishes that one is seeking a stationary point in a suitable subspace, and leads to the following characterisation.

Theorem 4.7.1. *For a given DMC, an input p.m.f.* $p = (p_x)_{x \in \mathcal{X}}$ *is such that* $I(X \wedge Y)$ *attains capacity iff there exists* C *such that*

$$\sum_y f(y|x) \log_+ \frac{f(y|x)}{\sum_u p_u f(y|u)} \begin{cases} = C, & \text{if } p_x > 0, \\ \leq C, & \text{if } p_x = 0; \end{cases}$$

C *is then the capacity.*

A method is still in general needed to find the p and C satisfying these conditions. The *Arimoto-Blahut algorithm* is a recursive scheme that does so. For any input p.m.f. p let

$$c_x(p) := \exp \sum_y f(y|x) \log_+ \frac{f(y|x)}{\sum_u p_u f(y|u)} \qquad (x \in \mathcal{X}).$$

Starting with $p^{(0)}$ any input p.m.f. with all components positive, the sequence $p^{(1)}, p^{(2)}, \ldots$ is defined successively by

$$p_x^{(k+1)} := \frac{p_x^{(k)} c_x(p^{(k)})}{\sum_u p_u^{(k)} c_u(p^{(k)})} \qquad (x \in \mathcal{X}).$$

With $X^{(k)}$ an input having p.m.f. $p^{(k)}$, and $Y^{(k)}$ the corresponding output, the sequence $I(X^{(k)} \wedge Y^{(k)})$ converges to C. Further, it is non-decreasing, so furnishes for any k an attainable reliable transmission rate, which can be made as close to C as desired by taking k large enough. See Blahut (1987), §5.4 for details.

Magnitude of the probability of error
Suppose, for a channel of capacity $C > 0$, we transmit at a fixed rate $r < C$. It is a consequence of our definitions in §3.1 that for large enough n-blocks there is a code γ_n with error probability $\beta(\gamma_n)$ as small as we wish. How fast does the best $\beta(\gamma_n)$ tend to zero as $n \to \infty$? It turns out that the convergence is *exponentially fast*, as fast as we could hope it could be. There is an exponent function $e^*(r) > 0$, specific to the channel, such that the lowest possible error-probability $\beta_{n,r}$ satisfies $\beta_{n,r} = e^{-n(e^*(r) + o(1))}$ as $n \to \infty$. Proving that much is not too hard; indeed for the binary symmetric channel you have shown in Exercise 3.5.4 that the minimal error-probability decays exponentially fast. However, finding $e^*(r)$ exactly is a hard problem, still not fully solved even for the BSC.

The other side of the coin is when we transmit at fixed rate r *above* the capacity C of the channel. Our definition of capacity now implies that the optimal error-probability $\beta_{n,r}$ for n-blocks cannot tend to zero as $n \to \infty$. How does it behave? With a sort of even-handed justice, its behaviour is about as bad as could be envisaged: it tends to 1 exponentially fast. So there is high probability of error somewhere in decoding the received n-block. Indeed, there is even a positive probability of *symbol* error in this situation. Rather than attempt to transmit at rate above capacity, one should encode the source down to rate C with a suitable distortion-measure, as in §2.13, and then transmit at capacity rate.

Details of the above will be found in Blahut (1987), §§5.7, 6.1.

Channels with input costs

Channel coding subject to a cost schedule closely parallels source coding under a fidelity criterion, as discussed in §2.13. We return to the set-up of §3.1 but assume now that every possible codeword $x \in C_n$ has a cost $c_n(x)$ that you must pay if you transmit it through the channel. The collection of all cost-functions c_n is the cost schedule. Typically, C_n is the set $\mathcal{X}^n$ of n-strings of letters from some fixed alphabet $\mathcal{X}$, there is a cost $c(x)$ associated with use of each letter $x \in \mathcal{X}$, and the cost of using a word is the sum of the costs of its letters:

$$c_n(x^{(n)}) = \sum_1^n c(x_j) \qquad (x^{(n)} = x_1 x_2 \ldots x_n \in \mathcal{X}^n). \tag{1}$$

Recall that we apply a coding γ_n to the random source-word $U^{(n)}$ which is assumed equidistributed over a set of $m = 2^{n(r+o(1))}$ values. The code-word generated is $X^{(n)} = \gamma_n(U^{(n)})$. Its cost is $c_n(X^{(n)})$, its expected cost is $Ec_n(X^{(n)}) = Ec_n(\gamma_n(U^{(n)}))$, and its per-letter expected cost is $\frac{1}{n}Ec_n(X^{(n)})$. We insist that the coding γ_n is such that this quantity is no higher than some pre-specified amount α. We say that the channel can transmit at rate r *reliably at cost-rate* α if it can transmit reliably at rate r using codings γ_n satisfying this extra restriction. The *capacity-cost function* $C(\cdot)$ of the channel is then defined by letting $C(\alpha)$ be the supremum of reliable transmission rates at cost rate α.

Analogously to the conclusion of the Second Coding Theorem, $C(\alpha)$ turns out to be given in general circumstances by the same formula as C,

$$C(\alpha) = \lim_{n \to \infty} \sup_{p_{X^{(n)}}} \frac{1}{n} I(X^{(n)} \wedge Y^{(n)}), \tag{2}$$

the supremum being now over the set of input p.m.f.s $p_{X^{(n)}}$ such that $Ec_n(X^{(n)}) \leq \alpha$. The capacity-cost function is defined on some interval $[\alpha_{\min}, \infty)$, is non-decreasing and can be shown to be concave.

You will observe that our treatment of the memoryless Gaussian channel can be fitted into the above framework. The constraints (C1) and (C2) of §3.6 are cost schedules, and our conclusions are evaluations and bounds of the resulting capacity-cost functions.

For a DMC the evaluation (2) simplifies to

$$C(\alpha) = \sup_{p_X} I(X \wedge Y)$$

where the supremum is over all p.m.f.s for the input letter X such that $Ec(X) \leq \alpha$. The input distribution for which channel capacity is attained must satisfy the cost bound for some sufficiently high $\alpha = \alpha_0$, say, and it then is plain that $C(\alpha)$ is constant for $\alpha \geq \alpha_0$.

Example 4.7.2. Consider a BSC with error probability p. The source alphabet is 0, 1 with costs $c(0) = 0$, $c(1) = 1$. The capacity-cost function evaluates as

$$C(\alpha) = \begin{cases} g\big((1-\alpha)(1-p) + \alpha p\big) - g(p), \\ \qquad\qquad \text{for } 0 \leq \alpha \leq \tfrac{1}{2}, \\ 1 - g(p), \\ \qquad\qquad \text{for } \alpha \geq \tfrac{1}{2}, \end{cases}$$

where $g(t) = -t \log t - (1-t)\log(1-t)$, and is illustrated in Fig. 4.7.1 for the case $p = \tfrac{1}{10}$.

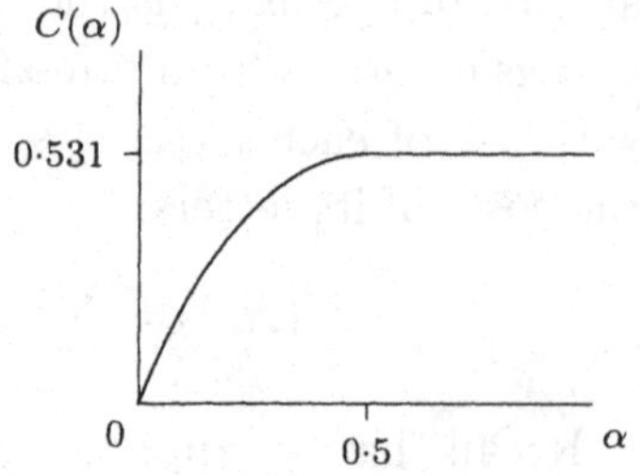

For further reading consult, for instance, McEliece (1984), §I.2.1.

Figure 4.7.1

Inequalities

From the text so far you will have the impression that only one inequality is needed for virtually all of the information-theory part of our subject. For most purposes that is the case, but one deep and important inequality holds for continuous entropy h_c. We give it for densities on $\mathbb{R}^d$, for which the notion of continuous entropy is treated in Exercise 4.5.4.

Entropy-power Inequality. *If X and Y are independent random n-vectors with densities, then*

$$e^{\frac{2}{n} h_c(X+Y)} \geq e^{\frac{2}{n} h_c(X)} + e^{\frac{2}{n} h_c(Y)},$$

with equality iff X and Y are Gaussian (normal) with covariance matrices proportional to each other.

The specialization to $\mathbb{R}$ thus says that, for independent r.v.s X, Y with densities, $e^{2h_c(X+Y)} \geq e^{2h_c(X)} + e^{2h_c(Y)}$, with equality iff X and Y have normal distributions.

The inequality was found by Shannon and proved by A. J. Stam in his 1959 Ph. D. thesis. The proof hinges on an intimate connection between differential entropy and the notion of *Fisher information* which we introduced in §2.13. The published version of the original proof, Stam (1959), is readable and worth consulting.

Many algebraic inequalities *consequent* on information theory are derived in Cover & Thomas (1988).

Other channels and systems

We have discussed only the simplest channels, but they are representative of more general arrangements. For instance, the channel can operate in continuous time as well as space; one then approximates it by a time-discretized version which is the Gaussian channel we have studied. Then, channels with memory have been extensively investigated. Our channels are 'two-terminal' — input and output — but more general many-terminal channels are worth modelling. There are diverse possibilities for how the terminals are linked and what the various users are permitted to do (e.g. receive only, or receive and transmit). Phrases such as the *broadcast channel, multi-user networks, multiple-access channel* will give some idea of the range. Advanced texts such as Ash (1965), Blahut (1987), Csiszár & Körner (1981), Gray (1990b), and papers such as Cover (1972), El Gamal & Cover (1980), and those in Longo (1977), are places to start to find out about these.

4.8 Problems

1. For any r.v.s X_1, X_2, X_3 define
    $$I(X_1 \wedge X_2 | X_3) := h(X_1 | X_3) + h(X_2 | X_3) - h(X_1, X_2 | X_3)$$
    ('conditional mutual information').

    (a) Show that
    $$I(X, Y \wedge Z) = I(Y \wedge Z) + I(X \wedge Z | Y)$$
    (recalling that the left-hand side is to be read as $I((X, Y) \wedge Z)$). This says that the information contained in (X, Y) about Z equals the information in Y about Z plus the information provided by X about Z in the knowledge of Y.

    (b) Deduce from Exercise 2.10.3 that $I(X \wedge Z | Y) \geq 0$, with equality if and only if X and Z are conditionally independent given Y.

2. *(Continuation)* Conditioning does not increase entropy (Theorem 2.10.2). However no such property exists for mutual information. Show that neither of $I(X \wedge Y)$ or $I(X \wedge Y | Z)$ dominates the other in general by finding counterexamples as follows:

    (a) jointly distributed binary r.v.s such that $I(X \wedge Y) = 0$ and $I(X \wedge Y | Z) \neq 0$;

    (b) jointly distributed r.v.s such that $I(X \wedge Y | Z) = 0$ and $I(X \wedge Y) \neq 0$.

    (Another way of putting this is that independence of X and Y neither implies nor is implied by their conditional independence given Z.)

3.   Let X, Y be jointly distributed discrete random variables. A message X is transmitted through a first channel, and the output Y of this channel is then transmitted through a second channel to give a final output Z. Thus the discrete random variables X, Y, Z are such that X and Z are independent, conditional on the value of Y. Show that

$$h(X|Y) \leq h(X|Z),$$

and determine conditions under which there is equality in this relation. (Cambridge 1981)

4.   Consider a pair of DMCs in cascade (see Exercise 4.3.1) and let X denote an input letter to the first component channel, Y the corresponding output used as input to the second component channel, and Z the output letter from the cascade.

(a) By applying Problem 3, or otherwise, show that

$$I(X \wedge Z) \leq I(Y \wedge Z),$$

with equality iff X and Y are conditionally independent given Z.

     This is known as the *Data-processing Lemma*. Roughly, 'mutual information cannot be increased by processing'.

(b) The capacity C of the cascade DMC is $C = \sup I(X \wedge Z)$, the supremum being over all distributions for X. Deduce from (a) that it is related to the capacities C_1, C_2 of the component DMCs by

$$C \leq \min(C_1, C_2).$$

(c) The equality condition in (a) may be put as that the conditional distributions $p_{X|Y,Z}$ and $p_{X|Z}$ are the same, so that in decoding (guessing X from knowledge of Z) it makes no difference if you are also told Y. Verify that this occurs
     (i) if the second channel is lossless,
     (ii) if the first channel is useless, and
     (iii) in other cases as well (consider the BSC).

(d) True or false? "When the output from the first channel is *recoded*, in blocks if need be, to optimize it as input to the second channel, the compound channel does have capacity $\min(C_1, C_2)$."

5.   When output letters for a DMC are combined into a single letter a *reduced channel* is said to result. Reduction is equivalent to adding a deterministic channel in cascade after the given DMC.

(a) Reduction clearly cannot increase channel capacity, and may lessen it. Prove that it leaves the capacity unaltered if and only if those columns of the channel matrix, that correspond to the letters to be combined into a single letter, are *proportional* to one another.

Hint: by Problem 4(a), you want X and Y conditionally independent given Z, which can be expressed as

$$P(X = x|Y = y, Z = z) = P(X = x|Z = z) = P(X = x|Y = y', Z = z).$$

Thus if y and y' are output letters to be combined into a single letter z you want

$$P(X = x|Y = y) = P(X = x|Y = y').$$

But for fixed y the numbers $P(X = x|Y = y)$, as x runs through the input alphabet, are proportional to the yth column of the channel matrix.

(b)Show that the DMC with channel matrix

$$\begin{pmatrix} \frac{5}{32} & \frac{3}{8} & \frac{5}{32} & \frac{5}{16} \\ \frac{7}{32} & \frac{1}{8} & \frac{7}{32} & \frac{7}{16} \end{pmatrix}$$

can be reduced to a binary channel with no loss of capacity.

6. Consider a lossless channel with binary (0/1) input and output, but with input sequences constrained by the condition that no symbol should occur in a run of length greater than two. Classify a binary sequence by its 'state' σ, the value (00, 01, 10, or 11) of the pair of symbols with which it terminates. Let $b_{\sigma\sigma'}$ be the number of ways in which the state of the sequence can change from σ to σ' by the addition of a permitted symbol, and define the matrix $B = (b_{\sigma\sigma'})$. What is B in this case?

   Show that if a sequence of length n must follow on from a sequence in state σ and end in state σ' then the number of such sequences is the $(\sigma\sigma')^{\text{th}}$ element of B^n. Hence show that the capacity of the channel is $C = \log(\frac{1}{2}(1 + \sqrt{5}))$. (Cambridge Dipl. Stat. 1988)
   (Problem 2.14.4(b) is similar.)

7. A spy sends messages to his contact as follows. Each hour either he does not telephone, or he telephones and allows the telephone to ring a certain number of times — not more than k, for fear of detection. His contact does not answer, but merely notes whether or not the telephone rings, and, if so, how many times. Because of deficiencies in the telephone system, calls may fail to be properly connected; correct connection has probability p, where $0 < p < 1$, and is independent for distinct calls, but the spy has no means of knowing which calls reach his contact. If connection is made, then

the number of rings is transmitted correctly. The probability of a false connection when no call is made may be neglected. Write down the channel matrix for this channel and calculate the capacity explicitly. Determine a condition on k in terms of p which will imply, with optimal coding, that the spy will always telephone. (Cambridge 1975)

8.   A memoryless discrete-time channel produces outputs Y from non-negative integer-valued inputs X by

$$Y = \varepsilon X,$$

where ε is independent of X, $P(\varepsilon = 1) = p$, $P(\varepsilon = 0) = 1 - p$ and inputs are restricted by the condition that $EX \leq 1$.

By considering input distributions $(a_i)_{i=0,1,\dots}$ of the form

$$a_i = \begin{cases} a, & \text{for } i = 0, \\ cd^i, & \text{for } i = 1, 2, \dots, \end{cases}$$

or otherwise, derive the optimal input distribution and determine an expression for the capacity of the channel. (Cambridge 1985)

9.   An input alphabet to a DMC has three letters 1, 2 and 3. The letter j is received as $j - 1$ with probability p, as $j + 1$ with probability p, and as j with probability $1 - 2p$, the letters of the output alphabet thus ranging from 0 to 4. Determine the form of the optimal input distribution, for general p, as explicitly as possible. Compute the channel capacity in the three cases $p = 0$, $p = \frac{1}{3}$ and $p = \frac{1}{2}$. (Cambridge 1983)

10.   The input and output of a memoryless channel both take values in the set {dot, dash, blank} and the channel matrix is

$$\begin{pmatrix} (1-r)p & (1-r)(1-p) & r \\ (1-r)(1-p) & (1-r)p & r \\ 0 & 0 & 1 \end{pmatrix}.$$

Determine the capacity of the channel. (Cambridge 1981)

11.   The *binary error-and-erasure channel*. A memoryless discrete-time channel may have input 0 or 1 at each instant, and its output is 0, 1 or θ where θ means 'unintelligible'. The probabilities that a transmitted symbol is received correctly, incorrectly, and unintelligibly are p, q, r respectively (where $p, q, r \geq 0$ and $p + q + r = 1$). Determine the optimal coding for, and the capacity of, the channel. (Cambridge 1971)

12.   Suppose, for a symmetric memoryless channel, that the input and output alphabets coincide, have r symbols, and that the probability that a symbol is incorrectly received is a given number $\varepsilon \in (0,1)$. Find bounds on the capacity of the channel, and exhibit cases where they are attained. (Cambridge 1989)

13.   An Indian wishes to send smoke signals. The signal is coded in puffs of smoke of different lengths: short, medium and long. One puff is sent per unit time. Assume a puff is observed correctly with probability p, and, with probabilities $1 - p$,
       (a) a short puff appears to be medium to the recipient,
       (b) a medium puff appears to be long,
       (c) a long puff appears to be short.
      What is the maximum rate at which the Indian can transmit reliably, assuming the recipient knows the encoding system he uses?
      It would be more reasonable to assume that a long puff may disperse completely rather than appear short, so the communication channel would no longer be symmetric. State briefly in what way this would affect your derivation of a formula for channel capacity. (Cambridge 1982)

14.   The text of an English-language message passes through a memoryless channel which transmits letters faithfully with probability $1 - \theta$ but which with probability θ transmits a letter mistakenly as the letter that follows it in the alphabetical order $abc\ldots xyz$ (mistaken transmission of z being recorded as a).
      Derive an expression for the capacity of the channel. (Cambridge 1974)

15.   A channel transmits two symbols; the output is corrupted by noise which affects at most one out of each block of b symbols. The probabilities of no error in a block, or of an error in the ith symbol of a block, are all $1/(b+1)$, independently of what happens to the other blocks. Find the capacity of the channel.
      Hint: Model the block transmission as a symmetric channel on 2^b symbols.

16.   Show that the geometric distribution on $\mathbb{Z}_+$ (the non-negative integers) has maximum entropy amongst all distributions on $\mathbb{Z}_+$ with the same mean.
      (cf. Problem 2.14.18)

Suppose that two non-negative integer-valued random variables X and Y are related by

$$Y = X + \nu$$

where the noise variable ν is independent of X and is geometrically distributed on $\mathbb{Z}_+$. Determine the distribution of Y that maximizes the mutual information of X and Y under the constraint that $E(X) \leq K$ and show that this distribution can be realized by assigning to X the value zero with a certain probability and letting it follow a geometric distribution (having an appropriate expectation) with the complementary probability. (Cambridge 1979)

17.   A random variable Y is distributed on the non-negative integers. Show that the maximum entropy of Y, subject to $E(Y) \leq m$, is $-m \log m + (m+1) \log(m+1)$, attained by a geometric distribution (by which is meant pq^j, for $j = 0, 1, \ldots$) with mean m.

A memoryless channel produces outputs Y from non-negative integer-valued inputs X by

$$Y = X + \varepsilon,$$

where ε is independent of X, $P(\varepsilon = 1) = p$, $P(\varepsilon = 0) = 1 - p = q$ and inputs X are constrained by $E(X) \leq q$. Show that, provided $p \leq \frac{1}{3}$, the optimal input distribution is

$$P(X = x) = \frac{1}{1+p} \left(\left(\frac{1}{2}\right)^{x+1} - \left(\frac{-p}{q}\right)^{x+1} \right) \qquad (x = 0, 1, 2, \ldots),$$

and determine the capacity of the channel.

Describe, very briefly, the problem of determining channel capacity if $p > \frac{1}{3}$. (Cambridge 1984)

18.   A channel consists of r independent memoryless Gaussian channels, the (additive) noise in the ith channel having variance v_i, for $i = 1, 2, \ldots, r$. The compound channel is subject to an overall power-constraint $E \sum_i X_{i,t}^2 \leq p$ for each t, where $X_{i,t}$ is the input to channel i at time t. The output from channel i at time t is $Y_{i,t} = X_{i,t} + \varepsilon_{i,t}$ where $\varepsilon_{i,t}$ is the noise. Determine the capacity of the compound channel. (Cambridge 1986)

Hint: if the power in the ith channel were restricted to p_i the capacity would be $C(p_1, \ldots, p_r) = \sum_i \frac{1}{2} \log(1 + p_i/v_i)$. Argue that the actual capacity is the maximum of C subject to $\sum_i p_i = p$. The required p_i are $p_i := \max(0, (2\lambda^*)^{-1} - v_i)$ where λ^* is the solution of $\sum_i \max(0, (2\lambda)^{-1} - v_i) = p$.

19.   Prove that the exponential distribution of mean μ has the great-

est continuous entropy among all continuous distributions on $(0, \infty)$ with means at most μ.

Consider a memoryless channel with additive exponential noise: the output Y_t from input X_t is

$$Y_t = X_t + \varepsilon_t$$

where X_t and ε_t are independent, and ε_t is exponentially distributed with mean 1. The input X_t may have any distribution on $[0, \infty)$ with mean at most K. Write down a formula for the capacity of a channel with additive noise. Show that it is possible to choose the distribution of X_1 so that Y_1 has maximal continuous entropy for a distribution with mean at most K. Deduce the capacity of the channel. (Cambridge Dipl. Stat. 1990)

ERROR-CONTROL CODES

5.1 Error-control codes

We retain the usual model from Chapter 0 of a transmission channel: the *sender* transmits some message to the *receiver* across a channel which may introduce errors. We are interested in this section in devising ways in which the message may be transmitted in order to ensure that a received message which has been corrupted may nonetheless be interpreted correctly. We shall be thinking of errors as relatively unlikely.

Definition. *A* code *C* *of length* *n* *and size* *m* *over an alphabet* *F* *with* $q = \#F$ *letters is a subset of* *m* *elements (*codewords*) of the set* $R = F^n$ *of possible words.*

In Chapter 1 we defined a code to be a map f from a set of m messages to the set of finite strings over an alphabet. However in Chapter 3 we identified a code with the image of the map f, that is, with the set of codewords: this is the point of view we are going to continue with in this chapter. We restrict our attention to block codes of fixed length n.

We say C has *parameters* $[n, m]$. The sender *encodes* the message by assigning elements of C to the m message symbols and transmitting the corresponding codeword. At the other end of the channel, possibly after errors have been introduced, words arrive which may not be codewords but in general simply elements of R. The receiver has to *decode* the received word to recover the original message.

We shall consider only codes of fixed length. The alphabet F could be any finite set, but in practice the most common class is that of the *binary* codes, those using the alphabet $F = \{0, 1\}$.

The designer of error-control codes has to balance a number of factors: the coder and decoder should be easy to build and fast to operate, the process should allow messages to be read even after errors, and the rate of transmission of information should not be reduced intolerably. The balance between these conflicting requirements will depend on the particular application.

Definitions. *A code C is d-error-detecting if making up to d changes in a codeword produces a result which can be detected as invalid: that is, making up to d changes in a codeword cannot produce another element of C. Call C e-error-correcting if making up to e changes results in an element of R from which the original codeword can be unambiguously reconstructed.*

Definition. *The information rate of C is*
$$\rho(C) = \frac{1}{n} \log_q m = \frac{\log m}{n \log q}.$$

We have $m = q^{n\rho(C)}$, so sending a codeword in time n is equivalent to sending $\rho(C)$ letters per unit time from an alphabet of size q. We made a similar point at the beginning of §2.2.

Example 5.1.1. The trivial $[n, q^n]$ code. We take the code C to be the whole of R, so that every possible message is a valid codeword. The information rate is 1, at the expense of having no error detection or correction.

Example 5.1.2. Parity check. A common error-detection system is to add a parity digit or check digit onto the end of a word. For example, eight-track paper tape and its electronic descendants would use the alphabet $F = \{0, 1\}$, $n = 8$ and $m = 2^7$. A digit 1 is represented by a hole punched in the paper tape, or by an electric potential. The message is encoded onto bits 1 to 7 and then an eighth bit is added so that the total number of bits equal to 1 is even. This will detect a single error, has information rate $\frac{7}{8}$ and the coding and decoding processes are simple. British Standard BS 6692 (1985) gives further details of the encoding.

Example 5.1.3. A more sophisticated version is found in the check digit added at the end of the International Standard Book Number, which is computed by addition modulo 11 over the alphabet $\{0, \ldots, 9, X\}$, where X denotes 10 mod 11. The book number has nine decimal digits and these are added with weights $10, 9, \ldots, 2$ to give the final check digit. The system is described by Standard Book Numbering Agency (1985).

ISBN is designed to detect single errors and *transpositions* of adjacent symbols, which are the commonest forms of error in typing. We shall not be dealing with transposition of symbols.

Example 5.1.4. The n-fold repetition $[n, q]$ code. The code formed by repeating each symbol n times is $(n - 1)$-error detecting and $\lfloor \frac{n-1}{2} \rfloor$-error correcting (why?). The information rate is $\rho = \frac{1}{n}$. The encoder is trivial; the decoder returns the 'majority verdict'. This is the code we used in Example 3.5.4 to show that naïve coding does not give a positive reliable transmission rate.

Recall that in §3.5 we defined the *Hamming distance* on F^n to be

$$\rho(x, y) = \# \{i \; : \; x_i \neq y_i\}.$$

Equivalently, you could define the distance between two vectors to be the least number of changes required to turn one into the other. If you did not do Exercise 3.5.2 at the time, here is the answer!

Proposition 5.1.5. *Hamming distance defines a metric on F^n. That is,*

(i) $\rho(x, y) \geq 0$ *and is* $= 0$ *iff* $x = y$;

(ii) $\rho(x, y) = \rho(y, x)$;

(iii)   *('Triangle inequality')*

$$\rho(x, y) \leq \rho(x, z) + \rho(z, y).$$

Proof. Parts (i) and (ii) are easy. For (iii), we see that to change x into y via z takes $\rho(x, z) + \rho(z, y)$ changes, and so this is at least the minimum number of changes required, which is $\rho(x, y)$.   □

The topology induced by this metric is the discrete topology, in which every set is open.

Let

$$B(x, r) := \{y \in F^n \; : \; \rho(x, y) \leq r\}$$

be the closed ball of Hamming radius r about x. The number of elements of F^n at Hamming distance exactly i from the point x is $\binom{n}{i}(q-1)^i$. since they must differ from x in exactly i coordinates and there are $q - 1$ possibilities in each such place. We see that $B(x,r)$ has

$$V_q(n,r) := \sum_0^r \binom{n}{i}(q-1)^i$$

elements.

Definitions. *The minimum distance $d = d(C)$ of the code C is the minimum Hamming distance between distinct elements of C. The error-control rate is d/n.*

We shall sometimes write the parameters of a code as $[n, m, d]$ where d is the minimum distance.

Proposition 5.1.6. *If C has minimum distance d then*

(i) C is $d - 1$-error detecting and
(ii) C is $\lfloor\frac{d-1}{2}\rfloor$-error correcting.

Proof. For (i), observe that $d - 1$ errors cannot change one codeword into another.

For (ii), note that the balls of radius $e = \lfloor\frac{d-1}{2}\rfloor$ about codewords must be disjoint: for if $x, y \in C$ and $z \in B(x,e) \cap B(y,e)$ then by the triangle inequality, $\rho(x,y) \leq \rho(x,z) + \rho(z,y) \leq e + e \leq d - 1$. $\qquad\square$

Hence to decode a received word in F^n we have to find the nearest element of C. This may not be easy. If you assume that up to e errors have occurred in the received word z then there are $V_q(n,e)$ elements which might be the transmitted word x. In the next section we shall introduce classes of codes for which there is a rule allowing you to reconstruct the transmitted word from the received word. Unfortunately this requirement tends to conflict with C having a high information rate.

Proposition 5.1.7: Hamming bound. *If C is an e-error correcting $[n, m]$ code then*

$$mV_q(n,e) = m\sum_{r=0}^e \binom{n}{r}(q-1)^r \leq q^n.$$

Proof. As in the proof of the previous Proposition, we see that the balls of radius e about the m codewords must be all be disjoint in pairs. Hence the total number of points in all these balls is at most the number of elements q^n in the message space. $\qquad\square$

Definition. *A code C is perfect if the bound of the Proposition is achieved, so that the e-balls centred on the codewords are disjoint and fill out F^n.*

The fact that the bound can be achieved by choice of the parameters q, n and m does not imply the existence of a code with those parameters. For example, although $\binom{90}{0} + \binom{90}{1} + \binom{90}{2} = 2^{12}$, there is no $[90, 2^{78}]$ perfect 2-error correcting binary code. (See Problem 1 at the end of the chapter.)

Example 5.1.8. Hamming's binary [7,16] code. This is a code of length 7 over the alphabet $\{0, 1\}$.

The codewords are the binary words $(x_1, \ldots, x_7)$ with the message encoded onto digits x_3, x_5, x_6, x_7 and parity bits x_1, x_2, x_4 satisfying parity checks

$$x_1 + x_3 + x_5 + x_7 = 0$$
$$x_2 + x_3 + x_6 + x_7 = 0$$
$$x_4 + x_5 + x_6 + x_7 = 0$$

(all taken modulo 2). To decode a received word $(z_1, \ldots, z_7)$, assumed to have at most one error, form the *syndrome* $s = (s_4, s_2, s_1)$ by

$$s_1 = z_1 + z_3 + z_5 + z_7$$
$$s_2 = z_2 + z_3 + z_6 + z_7$$
$$s_4 = z_4 + z_5 + z_6 + z_7$$

(again all taken modulo 2). If $s = (0, 0, 0)$ then the received word is correct. If not, then s interpreted as a binary number gives the number of the bit in error. (Exercise 5.1.6.)

It is easy to verify that this code has minimum distance 3, and so is a perfect 1-error correcting code with information rate 4/7. We shall see a simple proof of this fact in §5.2. In fact this code has the highest information rate of any $[n, 16]$ 1-error correcting code.

It is said that Hamming invented this code after several attempts to read a paper tape into an early computer (using the parity check) failed and in exasperation he said "If the machine can *detect* an error, why can't it locate the position of the error and *correct* it?".

Exercises

1.   Computer memory chips often use only a single parity bit for error detection rather than a more sophisticated system of error correction. Why do you think this might be?

2.   Use the Hamming [7, 16] code to encode the messages 0000, 1100,

0011 and 1111. Decode the received messages 1000000, 0000001, 1111111 and 0010111.

3. List the codewords of the Hamming [7, 16] code and show that if the bits are written in the order $(x_5, x_7, x_6, x_3, x_4, x_2, x_1)$ then the cyclic shift of a codeword is again a codeword.

4. The n-fold repetition code is perfect iff n is odd.

5. Show that the results of Proposition 5.1.6 are sharp, that is, a code of minimum distance d cannot detect d errors or correct $\lfloor \frac{d+1}{2} \rfloor$.

6. Verify the rule for correcting single errors in the Hamming [7,16] code. Show that if two errors occur, the decoder always gives the wrong answer.

7. Let C be the binary [11, 12] code consisting of the word 10111000100 and its ten cyclic permutations, together with 00000000000. Show that C has minimum distance 5.

5.2 Linear codes

You may find it helpful to review Sections 1, 2 and 4 of the Appendix before starting this section.

When F is a finite field the message space F^n has a natural structure as a vector space over F. The most well-known examples of finite fields are the *prime fields* $\mathrm{GF}(p) = \mathbb{Z}/p$ of integers modulo a prime number p, and an important case is the class of binary codes with $p = 2$. In this section we shall be assuming that F is a finite field.

Definition. *A linear code of length n over a finite field F is a subspace of F^n regarded as a vector space over F. The rank of a linear code is its dimension r and we say that the code has parameters (n, r) (i.e. $[n, q^r]$) or (n, r, d) where d is the minimum distance.*

The information rate ρ of a linear code is just $\frac{r}{n}$.

The advantages of linear codes include ease of design and construction of the encoder and decoder. Disadvantages include the fact that not all codes are linear, so that the most efficient codes for a given set of parameters may turn out to be non-linear.

Definition. *A generator matrix for C is a $r \times n$ matrix G whose rows form a basis for C.*

By performing row operations on G we may put it into the *standard form* $G = (I_r | G_1)$ where I_r is a $r \times r$ identity matrix and G_1 is $r \times (n - r)$.

We see that a linear code C is *systematic*, that is, there is a subset of r of the coordinates $1 \ldots n$ (in fact, the first r) with the property that the elements of C uniquely correspond to these coordinates. Hence encoding is easy once we know the standard form of the generator matrix.

The parity and Hamming [7, 16] codes described in §1 are linear, because in each case the set of permissible codewords is defined by the linear relations on the words given by the parity conditions. We can identify the n-dimensional space of column vectors over F with the set of possible linear relations on the coordinates of the row vectors in F^n.

Definition. *The parity check for a linear code C is the space C° of parity check vectors which define C:*

$$C^\circ = \{p \in F^n \ : \ \forall x \in C, \quad xp = 0\}.$$

The parity check is the annihilator of C as defined in §7.4. It is a subspace of F^n of dimension $n - r$ where r is the dimension of C. Dually, given a subspace P of F^n we can define the corresponding code $C = P^\circ$ by

$$P^\circ = \{x \in F^n \ : \ \forall p \in P, \quad xp = 0\}.$$

C is a subspace of F^n of dimension $n - \dim P$ and we have the relations $(C^\circ)^\circ = C$, $(P^\circ)^\circ = P$.

Definition. *A parity check matrix H is a $n \times (n - r)$ matrix whose columns form a set of basis elements for $P = C^\circ$.*

By column operations we may bring a parity check matrix H into standard form

$$H = \left(\frac{H_1}{I_{n-r}}\right)$$

where I_{n-r} is an $(n - r) \times (n - r)$ identity matrix and H_1 is $r \times (n - r)$. Given a parity check matrix, C may be characterised as

$$C = \{x \in F^n \ : \ xH = 0\}.$$

Some other texts define the parity check matrix to be the transpose of our definition.

Example 5.2.1. The paper tape code is linear with generator matrix

$$\begin{pmatrix} 1 & 0 & 0 & 0 & 0 & 0 & 0 & 1 \\ 0 & 1 & 0 & 0 & 0 & 0 & 0 & 1 \\ 0 & 0 & 1 & 0 & 0 & 0 & 0 & 1 \\ 0 & 0 & 0 & 1 & 0 & 0 & 0 & 1 \\ 0 & 0 & 0 & 0 & 1 & 0 & 0 & 1 \\ 0 & 0 & 0 & 0 & 0 & 1 & 0 & 1 \\ 0 & 0 & 0 & 0 & 0 & 0 & 1 & 1 \end{pmatrix}.$$

and parity check matrix $(1,1,1,1,1,1,1,1)^\top$.

Example 5.2.2. The Hamming $[7,16]$ code is linear. Renumber the bits in the order $(x_3, x_5, x_6, x_7, x_1, x_2, x_4)$, so that the message bits come first; then a generator matrix is

$$\begin{pmatrix} 1 & 0 & 0 & 0 & 0 & 1 & 1 \\ 0 & 1 & 0 & 0 & 1 & 0 & 1 \\ 0 & 0 & 1 & 0 & 1 & 1 & 0 \\ 0 & 0 & 0 & 1 & 1 & 1 & 1 \end{pmatrix}$$

with a parity check matrix

$$\begin{pmatrix} 1 & 0 & 0 \\ 0 & 1 & 0 \\ 1 & 1 & 0 \\ 0 & 0 & 1 \\ 1 & 0 & 1 \\ 0 & 1 & 1 \\ 1 & 1 & 1 \end{pmatrix}.$$

The transpose of the parity check matrix for a code C is a $(n-r) \times n$ matrix, the generator for a code of length n and rank $n-r$. We call this code the *dual code* $C^\perp$ of C.

In the language of dual vector spaces, the space of possible parity check vectors and the space of possible words are dual spaces and ° is the annihilation operator. You may be happier with the following alternative description in terms of a scalar product on F^n.

We define the scalar product (x,y) on F^n by the usual Euclidean formula $(x,y) = \sum_{i=1}^n x_i y_i$. Note that the values are taken in F so that this is not an inner product in the usual sense. For example, the vector $(1,1) \in GF(2)^2$ is non-zero but has scalar product with itself $1.1+1.1$ which is 0 as an element of $GF(2)$. The orthogonal space

$$C^\perp = \{y \in F^n : \forall x \in C \quad (x,y) = 0\}$$

to C is also a subspace: the *dual* code to C. The dual of an (n, r) code is an $(n, n - r)$ code.

Definitions. *A parity check vector for a linear code C is the transpose of an element of $C^{\perp}$ and a parity check matrix H is the transpose of a generator matrix for $C^{\perp}$. We have*

$$C = \{x \in F^n \ : \ xH = 0\}.$$

Definition. *The weight $w(x)$ of a vector $x \in F^n$ is the number of non-zero places: $w(x) = \# \{i \ : \ x_i \neq 0\}$.*

Proposition 5.2.3.

(i) $w(x) \geq 0$;
(ii) $w(\lambda x) = w(x)$ for $\lambda \in F^*$;
(iii) $w(x + y) \leq w(x) + w(y)$.

Proof. The proof is immediate from Proposition 5.1.5.   □

Hence the minimum distance of a linear code is just the minimum non-zero weight.

Definition. *The weight enumerator polynomial of a linear code C is the homogeneous two-variable polynomial*

$$A(C, x, y) = \sum_{i=0}^{n} A_i x^i y^{n-i}$$

where A_i is the number of elements of C of weight i.

Since $A(C, x, y)$ is homogeneous, $A(C, x, y) = y^n A(C, x/y, 1)$, and so A can be regarded as a polynomial in the single variable x/y. Some authors take this approach.

Suppose that we are using a binary symmetric channel (§3.4), so that errors in transmission occur independently with probability p. The probability that a valid codeword is received is then $A(C, p, 1 - p)$ and this does not depend on the probability distribution on the codewords transmitted.

Theorem 5.2.4: MacWilliams identity.

$$A\left(C^{\perp}, x, y\right) = q^{-r} A(C, y - x, y + (q - 1)x)$$

We shall not give a proof of the general case of this theorem. See for example MacWilliams and Sloane (1977) Theorem 5.2.1 or van Lint (1986) Theorem 3.5.3. Problem 5.7.3 tackles the case $q = 2$.

In order to compute the weight enumerator of a linear code C of rank r, it would be necessary to find the weights of all q^r codewords. If, however, $n - r$ is small, then it may be easier to find the weights of the words of the dual code $C^\perp$ instead and then apply the identity. This is so, for example, in the case of the Hamming codes defined below: see Example 5.2.5 and Exercise 5.2.9.

Syndrome decoding

Definition. *If H is a parity check matrix for C, so that $x \in C \Leftrightarrow xH = 0$, then for a received message $z \in F^n$ we call zH the* syndrome *of z.*

In general we need to determine how to correct a received message z from the syndrome zH. The syndrome is constant on cosets $C + z$, so we need only tabulate the syndromes of a set of coset representatives. A useful representative to choose is a *leader*, a vector of least weight in the coset: this choice is not necessarily unique. For a (n, r) code the number of leaders we need to tabulate is

$$\frac{q^{n-r} - 1}{q - 1} + 1$$

and this is practical if $n - r$ is small, that is, if the rate r/n is close to 1.

Syndrome decoding is particularly simple for 1-error correcting codes. Assume that the received message $z = c + \lambda e_i$ where e_i is a coordinate vector and $c \in C$. Then $zH = \lambda e_i H$, so zH is a multiple of a row of H. The row number i determines which digit of x is in error. If C is binary this corrects the error immediately and otherwise the value of λ can be determined from any non-zero entry in the row e_i. We saw in Example 5.1.8 and Exercise 5.1.6 that the Hamming $[7, 16]$ code is particularly easy to decode using the syndrome.

Majority-logic decoding

Another method of decoding for linear codes is *majority-logic*.

Definition. *A set of $2e$ parity check vectors $y^{(1)}, \ldots, y^{(2e)}$ is* orthogonal *for position i if it satisfies*

(i) $y_i^{(m)} = 1$ *for each m and*
(ii)   *for each $j \neq i$ at most one of the $y^{(m)}$ has $y_j^{(m)} \neq 0$.*

Suppose that z is a received word containing at most e errors. If position i in z is correct, then $(z, y^{(m)})$ will be non-zero for at most e values of m, namely those for which position j is incorrect in z and $y_j^{(m)} \neq 0$. If position i in z is incorrect, then $(z, y^{(m)})$ will be non-zero for at least $2e - (e - 1) = e + 1$ values of j, as there can only be at most $e - 1$ possible errors which make $(z, y^{(m)})$ zero. We can thus determine whether position i is correct from the majority verdict of the values $(z, y^{(m)})$. (In case of equal votes, favour 'correct'.)

We can determine the correctness of each symbol in the message in turn if we can find a set of orthogonal check vectors for each position. Unfortunately, not every code can be decoded in this way (see Exercise 5.2.9).

Example 5.2.5. The Hamming $\left(\dfrac{q^d - 1}{q - 1}, \dfrac{q^d - 1}{q - 1} - d \right)$ code.

There are $n = \dfrac{q^d - 1}{q - 1}$ lines through the origin in a space D of dimension d over the field $\mathrm{GF}(q)$, since there are $q^d - 1$ non-zero points and $q - 1$ non-zero multiples of each point on each line. Choose one non-zero point on each line and write the coordinates as the rows of a $n \times d$ parity check matrix H. Any two rows of H are linearly independent, since they are coordinates of points on different lines in D.

Definition. *The $(n, n - d)$ Hamming code over $\mathrm{GF}(q)$ is the code C of length n and rank $n - d$ defined by the parity check matrix H.*

Any non-zero element of the Hamming code C corresponds to a linear relation on the rows of H. As any two rows of H are linearly independent, a linear relation among the rows of H must therefore involve at least three rows, and so each element of C has weight at least 3. Hence the Hamming code is 1-error correcting. It is easy to check that the code is perfect (Exercise 5.2.8). The Hamming code can thus be decoded easily by syndrome decoding.

The Hamming code over $\mathrm{GF}(2)$ with $d = 3$ is the $[7, 16]$ example in the previous section.

Exercises

1.   The n-fold repetition code is linear with parameters $(n, 1)$. Write down a generator and parity check matrix, determine the weight enumerator and identify the dual code. Find a set of orthogonal check vectors for each position.

2.   Find a standard form for the parity check matrix for the Hamming

(7, 4) code given in Example 5.1.8.

3.   What is the relation between the matrices G_1 and H_1 in the standard forms for a generator matrix and a parity check matrix?

4.   Assume C has odd minimum distance $d = 2e + 1$. Show that the vectors of weight $\leq e$ are all leaders in different cosets and that if C is perfect then all leaders are of this form.

5.   Let $A(C, x, y)$ be the weight enumerator of the linear (n, r) code C. Show that $A(C, 0, 1) = 1$ and $A(C, 1, 1) = q^r$. If C is binary, show further that $A(C, 1, 0) = 0$ or 1 and that $A(C, x, y)$ is symmetric in x and y if and only if $A(C, 1, 0) = 1$.

6.   What are the weight enumerators of the repetition (n, q) code and the trivial (n, q^n) code?

7.   List the codewords of the Hamming $(7, 4)$ code and show that it has weight enumerator $y^7 + 7x^3y^4 + 7x^4y^3 + x^7$. Verify the MacWilliams identity for this code.

8.   Verify that the binary Hamming $(n, n - d)$ code, $n = 2^d - 1$, is perfect.

9.   Show that every non-zero parity check vector for the binary Hamming $(n, n - d)$ code has weight 2^{d-1}. What does this imply about the majority-logic decoding for this code? Obtain the weight enumerator of the Hamming code.

10.   Generalise the 'binary number' trick for the Hamming $[7, 16]$ code given in Example 5.1.8 to other binary Hamming codes.

11.   The binary $[4, 4]$ code with words 0001, 0010, 0100, 1000 is not systematic and hence not a linear $(4, 2)$ code.

12.   Show that two applications of the MacWilliams identity do indeed return the original weight enumerator.

5.3   Constructing codes from other codes

Definition. *For a code C of length n over $\mathrm{GF}(q)$, the (parity check digit) extension C^+ is the length $n + 1$ code*

$$C^+ = \left\{ (x_0, x_1, \ldots, x_n) \ : \ (x_1, \ldots, x_n) \in C, \ \sum_{0}^{n} x_i = 0 \right\}.$$

This operation clearly does not change the size of C. If C is linear then so is C^+.

If C is binary (i.e. $q = 2$) with an odd minimum distance d then C^+ has minimum distance $d + 1$, since the weights of elements of C^+ are all even. The traditional parity check is the extension of the trivial (n, n) code.

Definition. *The* truncation (puncturing) *of a code C of length n is the code C^- of length $n - 1$ obtained by omitting the last symbol of each word of C.*

If C is linear then so is C^-.

Proposition 5.3.1. *Let C be a code of length n, size m and minimum distance $d \geq 2$. The truncation C^- is a code of length $n - 1$, size m and minimum distance d or $d - 1$.*

Proof. Clearly C^- has length $n - 1$. The size of C^- will be the same as that of C unless there are two words of C which differ only in the last digit (in which case they would have the same truncation). If so, such a pair of words would be distance 1 apart, but the minimum distance $d(C) \geq 2$. Finally, the distance, between codewords, the number of places in which they differ, is reduced by at worst 1 on truncation.   □

Definition. *The* shortening *of a code C of length n is the code C' obtained by taking the subcode of C consisting of all elements ending in a fixed symbol and then truncating this sub-code.*

Shortening decreases the length and size but does not change the minimum distance. Note that if C is linear and the chosen symbol is not 0, then C' is not necessarily linear.
Note that truncation and shortening are not the same thing!

Definition. *The* repetition $C^{(m)}$ *of a code C is the code of length mn obtained by repeating every word of C m times.*

If C is linear then so is $C^{(m)}$. The minimum distance of $C^{(m)}$ is clearly $m\, d(C)$. The m-fold repetition code is the repetition of the trivial $[1, q]$ code.

A natural construction from a pair of codes might be to consider the direct product $C_1 \times C_2$ but this is usually of little interest since the minimum distance of the product is the smaller of the minimum distance of the factors. For linear codes there is a related construction of some value.

Definition. *Let C_1 and C_2 be linear codes of length n with C_2 a subset of C_1. The bar product $C_1|C_2$ is the code of length $2n$ with words $(x|x + y)$ where $x \in C_1$ and $y \in C_2$.*

The $|$ symbol here has nothing to do with conditional probabilities.

Proposition 5.3.2. *The bar product of C_1 and C_2 is a linear code of length $2n$, rank $r(C_1) + r(C_2)$ and minimum weight at least $\min\{2d(C_1), d(C_2)\}$.*

Proof. Put $z = (x|x + y)$. We have $w(z) = w(x) + w(x + y)$. Suppose that $z \neq 0$, so that x and $x + y$ are not both zero. If $x = 0$ then $y \neq 0$ and $w(z) = w(y) \geq d(C_2)$. If $x \neq 0$ and $y = -x$ then $w(z) = w(x) \geq d(C_2)$ since in this case $x = -y$ must be in C_2. If $x \neq 0$ and $y \neq -x$ then $x + y \in C_1$ and so $w(z) = w(x) + w(x + y) \geq d(C_1) + d(C_1)$ as required. $\square$

The bar product can be decoded easily if decoders for C_1 and C_2 are available. Suppose that $(x|x + y)$ is transmitted and $(u|v)$ is received, possibly with e errors, where $e < \frac{1}{2}d(C_1|C_2)$. Since $e < \frac{1}{2}d(C_2)$, we can decode $v - u$ in C_2 to obtain y, even if all e errors are in the second part of the word. One of u and $v - y$ has $\leq \frac{1}{2}e < d(C_1)$ errors, so one of these can be decoded in C_1. Pass each through the decoder and choose the output which gives the correct value for x, that is, the value of x for which $(x|x + y)$ encodes to $(u|v)$.

Exercises

1.   The shortening of a linear code with respect to the symbol 0 is again linear. Is the shortening with respect to any other symbol linear?

2.   It is possible to shorten any code without decreasing the information rate.

3.   Write down a generator matrix for the parity extension of the Hamming $(7, 4)$ code.

4.   If C_1 has generator matrix G_1 and C_2 has generator matrix G_2, write down a generator matrix for $C_1|C_2$.

5.4 Bounds on error-control codes

Clearly we would like to design our codes so that the information rate and error-control rate are both large, and these requirements are pulling in opposite directions. In this section we discuss some of the consequent constraints on the design of a code. An example which you have already seen is the Hamming bound, Proposition 5.1.7, on the size of a code with given error-detection capability.

Let $A_q(n, d)$ denote the maximum size of a code of length n with minimum distance d over an alphabet of q symbols. Recall from §1 that we defined $V_q(n, d)$ to be the number of points in the closed Hamming ball of radius d in F^n.

We repeat the Hamming bound to show how it fits into this family of results.

Theorem 5.4.1. *If $d = 2e + 1$, then*

$$A_q(n, d) \leq q^n / V_q(n, e).$$

□

There is another simple upper bound for $A_q(n, d)$.

Theorem 5.4.2: Singleton bound.

$$A_q(n, d) \leq q^{n-d+1}.$$

Proof. Consider truncating C. Since truncation reduces the minimum distance by at most one (Proposition 5.3.1), we can truncate $d - 1$ times to produce a code in F^{n-d+1} with the same size m as C. But then $m \leq q^{n-d+1}$.
□

We derived the Hamming bound by trying to pack the Hamming balls into the space F^n. If instead we consider covering the space then we obtain a lower bound for $A_q(n, d)$.

Theorem 5.4.3: Gilbert–Varshamov bound.

$$A_q(n, d) \geq q^n / V_q(n, d - 1).$$

Proof. Suppose that C is a code of maximal size $m = A_q(n, d)$ in F^n; that is, no word can be added to C without decreasing the minimum distance $d = d(C)$. But then every word $z \in F^n$ must be distant at most $d - 1$ from some codeword: otherwise, if z were distance at least d from every word in C, we could add z to C and $C \cup \{z\}$ would still have minimum distance d.

So the balls of radius $d - 1$ round the m codewords must cover the whole space F^n. Hence $m V_q(n, d - 1)$ counts the words in F^n at least once each, and so is at least q^n.   □

To compare these results for large n, we need to find the power of q which best approximates the function $V_q(n, d)$.

We use the O, o notation as in §§2.2, 2.7: $O(f(n))$ denotes any function g (not necessarily the same on each occasion) such that $\frac{g(n)}{f(n)}$ is bounded as $n \to \infty$ and $o(f(n))$ similarly denotes any function g such that $\frac{g(n)}{f(n)} \to 0$ as $n \to \infty$.

Lemma 5.4.4.

(i)
$$\ln n! = n \ln n - n + O(\ln n);$$

(ii)
$$\log \binom{n}{r} = -(n - r) \log \left(\frac{n - r}{n} \right) - r \log \left(\frac{r}{n} \right) + O(\log n)$$
for any values of r with $0 \le r \le n$.

Proof. Part (i) is a weak form of Stirling's formula and is quite standard: see, for example, Abramowitz and Stegun (1975) (6.1.38) or Feller (1970) II.9. Part (ii) follows immediately by substituting (i) for $\log n!$ in $\log \binom{n}{r}$. $\square$

Note that in (i) we require natural logarithms, whereas in (ii) we can take any base, scaling by an appropriate factor.

For fixed δ, $0 < \delta < 1$, put
$$\alpha(\delta) := \limsup \frac{1}{n} \log_q A_q(n, \delta n).$$
You could think of $n\alpha$ as the largest power of q which can be used to approximate $A_q(n, \delta n)$. We shall call an estimate for $\alpha(\delta)$ an *asymptotic bound*.

Definition. *The entropy function $H_q(x)$ on $[0, \theta]$, where $\theta = 1 - 1/q$, is given by*
$$H_q(x) = \begin{cases} 0, & \text{for } x = 0; \\ x \log_q(q - 1) - x \log_q x - (1 - x) \log_q(1 - x), & \text{for } 0 < x \le \theta. \end{cases}$$

If $q = 2$ this reduces to the formula for the entropy of a Bernoulli source with probabilities x and $1 - x$ (§1.5) and also its information rate H (Theorem 2.5.1).

Proposition 5.4.5. *Let $0 \le \lambda \le \theta = 1 - 1/q$, $q \ge 2$. Then*
$$\lim_{n \to \infty} \frac{1}{n} \log_q V_q(n, \lambda n) = H_q(\lambda).$$

Proof. By definition, we have

$$V_q(n, \lambda n) = \sum_{i=0}^{m} \binom{n}{i}(q-1)^i$$

where $m = \lfloor \lambda n \rfloor$. We shall first show that the terms in this sum are increasing, so that the last term is the greatest.

We need to show that the ratio of successive terms

$$\frac{\binom{n}{i+1}(q-1)^{i+1}}{\binom{n}{i}(q-1)^i} = \frac{n-i}{i+1}(q-1)$$

is greater than 1 for $i + 1 \le m$. Now

$$i + 1 \le m \le \lambda n \le \theta n = \frac{n(q-1)}{q}$$

and you can check that the outer inequality

$$i < \frac{n(q-1)}{q}$$

is equivalent to

$$1 < \frac{n-i}{i+1}(q-1)$$

as claimed.

Now we have

$$\text{largest term} \le \text{sum} \le \text{number of terms} \times \text{largest term}$$

so that

$$\binom{n}{m}(q-1)^m \le V_q(n, m) \le (m+1)\binom{n}{m}(q-1)^m$$

and using the previous Lemma,

$$m \log(q-1) - m \log\left(\frac{m}{n}\right) - (n-m)\log\left(\frac{n-m}{n}\right) + \mathrm{O}(\log n)$$

$$\le \log V_q(n, m)$$

$$\le m \log(q-1) - m \log\left(\frac{m}{n}\right) - (n-m)\log\left(\frac{n-m}{n}\right)$$

$$+ \log(m+1) + \mathrm{O}(\log n).$$

Dividing through by n, fixing q as the base of logarithms and using the fact that $\frac{m}{n} = \lambda + \mathrm{O}\left(\frac{1}{n}\right)$, we have

$$\lambda \log(q-1) - \lambda \log(\lambda) - (1-\lambda)\log(1-\lambda) + o(1)$$

$$\le \frac{1}{n} \log V_q(n, \lambda n)$$

$$\le o(1) + \lambda \log(q-1) - \lambda \log(\lambda) - (1-\lambda)\log(1-\lambda) + o(1)$$

where the first $o(1)$ in the upper bound arises from $\frac{1}{n}\log(1+m)$. But now both the upper and lower bounds for $\frac{1}{n}\log V_q(n, \lambda n)$ are of the form $H_q(\lambda) + o(1)$, so the result follows.   □

We can now use this Proposition to derive the asymptotic versions of the three bounds for $\alpha(\delta)$.

Theorem 5.4.6: Asymptotic Hamming bound.
$$\alpha(\delta) \leq 1 - H_q(\tfrac{1}{2}\delta).$$

□

Theorem 5.4.7: Asymptotic Singleton bound.
$$\alpha(\delta) \leq 1 - \delta.$$

□

Theorem 5.4.8: Asymptotic Gilbert–Varshamov bound.
$$\alpha(\delta) \geq 1 - H_q(\delta), \quad for \quad 0 \leq \delta \leq \theta.$$

□

Asymptotic bounds for $q = 2$  Asymptotic bounds for $q = 49$

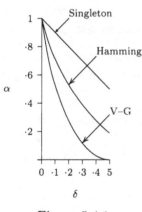
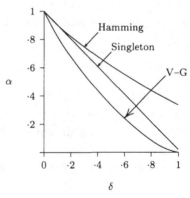

Figure 5.4.1   Figure 5.4.2

Exercises

1.   Compare the Hamming and Singleton bounds for various values of q.

Computer investigations
You will need a computer for these investigations. You should probably

represent each word in F^n as an array, length n, of integers in the range 0 to $q - 1$, and a code as a two-dimensional array with the codewords as the rows.

   The object of the investigations is to see how good randomly generated codes can be. In each case you should plot the information rate against the error-control rate on a graph and also mark in the various bounds from this section. Try for a range of values of q from 2 upwards, and up to as large a value of n as you can in each case. How good are your random codes?

2.   Write procedures to find the distance between two codewords and to find the minimum distance of a code.

3.   Write a program which generates successive codewords at random and finds the minimum distance of the code they form.

4.   Write a program which reads in the length n and required minimum distance d and generates random codewords, rejecting any which would make the minimum distance of the code less than d. Your program should stop when no more codewords have been accepted for a suitably long time.

5.   Write procedures to list all the elements of a linear code with a given set of generators and to find its minimum weight.

6.   Write a program which generates successive codewords at random and find the minimum weight of the linear code which they generate.

7.   Write a program which reads in the length n and required minimum weight d and generates random codewords, rejecting any which would make the minimum weight of the linear code which they generate less than d. Your program should stop when no more codewords have been accepted for a suitably long time.

5.5   Reed–Muller codes

Consider the points of d-dimensional space $D = \mathrm{GF}(2)^d$ listed in some order as $P_0, \ldots, P_{n-1}$ where $n = 2^d$. (You might read the coordinates off as binary numbers.) We make the row vectors $x = (x_0, \ldots, x_{n-1})$ in $R = \mathrm{GF}(2)^n$ correspond to subsets of D by the correspondence

$$x \leftrightarrow \{P_i \in D \ : \ x_i = 1\}.$$

We can interpret the row vectors as giving the values of the *character-*

istic or *indicator* functions on D. The *wedge product* operation $x \wedge y = (x_0 y_0, \ldots, x_{n-1} y_{n-1})$ then corresponds to the operation of intersection on subsets of D.

The operation $\wedge$ on binary vectors has nothing to do with the notation for mutual information introduced in Chapter 4.

In particular, the zero vector in R corresponds to the empty set as a subset of D and the vector $v_0 := (1, 1, \ldots, 1)$ corresponds to D as a subset of itself. Note that if x corresponds to $U \subseteq D$ then $v_0 + x$ corresponds to the complement $D \setminus U$ of U in D. The weight of an element $x \in R$ just gives the number of elements in the corresponding subset of D. The scalar product on R satisfies $(x, y) \equiv w(x \wedge y) \bmod 2$.

The operations $\wedge$ and $+$ turn R into a ring isomorphic to the power set $\mathcal{P}D$. See Example 7.1.3 and Problem 7.6.1

Let $v_1, \ldots, v_d$ denote the row vectors corresponding to the coordinate hyperplanes,

$$v_i \leftrightarrow H_i := \{ P \in D \ : \ i^{\text{th}} \text{coordinate of } P = 0 \} .$$

Let $H_i(0) := H_i$ denote the set of points with i^{th} coordinate 0 and $H_i(1) := D \setminus H_i$ its complement, the set of points with i^{th} coordinate 1. Then for $p = 0$ or 1, the vector $pv_0 + (1 - p)v_i \in R$ corresponds to $H_i(p)$. For distinct vectors $v_{i_1}, \ldots, v_{i_k}$, their wedge product has weight

$$w(v_{i_1} \wedge \ldots \wedge v_{i_k}) = 2^{d-k}.$$

We make the convention that a wedge product of zero terms denotes v_0.

Definition. *The* Reed–Muller code $RM(d, r)$ *of order r and length $n = 2^d$ is the subspace of the n-dimensional space* $\text{GF}(2)^n$ *spanned by v_0 and the $v_{i_1} \wedge \ldots \wedge v_{i_s}$ for $1 \le s \le r$ and $1 \le i_j \le d$.*

Example 5.5.1. The Reed–Muller codes of order 3. Order the elements of $D = \text{GF}(2)^3$ in 'binary number' order, so that $P_0 = 000$, $P_1 = 001$, ..., $P_7 = 111$. The subspace H_1 consists of all elements of D with 0 in the first co-ordinate, so $H_1 = \{P_0, P_1, P_2, P_3\}$. The generators of the various Reed–Muller codes of order 3 are given by the table overleaf.

Proposition 5.5.2. *The RM code of order r and length 2^d has rank $1 + \binom{d}{1} + \ldots + \binom{d}{r}$ and the $v_{i_1} \wedge \ldots \wedge v_{i_s}$ form a set of basis vectors for the code.*

Proof. We shall show that the $n = 2^d$ elements of the form $v_{i_1} \wedge \ldots \wedge v_{i_s}$ for $0 \le s \le d$ form a basis for $R = F^n$. If so, they are linearly independent and any subset of them must form a basis for the subspace they generate.

Table 5.5.1. The generators of $RM(3,d)$

	P_0	P_1	P_2	P_3	P_4	P_5	P_6	P_7
v_0	1	1	1	1	1	1	1	1
v_1	1	1	1	1	0	0	0	0
v_2	1	1	0	0	1	1	0	0
v_3	1	0	1	0	1	0	1	0
$v_1 \wedge v_2$	1	1	0	0	0	0	0	0
$v_1 \wedge v_3$	1	0	1	0	0	0	0	0
$v_2 \wedge v_3$	1	0	0	0	1	0	0	0
$v_1 \wedge v_2 \wedge v_3$	1	0	0	0	0	0	0	1

In turn, to show that they form a basis for the space R it is sufficient to show that they span the space, since there are just 2^n of them and this is the dimension.

Let e_j $(0 \leq j \leq n-1)$ be the vector in R with digit 1 in the jth position and 0 elsewhere: so the e_j form the standard basis of R. We shall show that the e_j can be expressed in terms of v_0 and products of the form $v_{i_1} \wedge \ldots \wedge v_{i_s}$.

We can regard e_j as the characteristic function of the set $\{P_j\}$ consisting of a single point P_j. The point P_j is determined by its coordinates $\pi_i(j)$ say, so that $\{P_j\} = \bigcap_{i=1}^d H_i(\pi_i(j))$. Hence we can write the corresponding formula

$$e_j = \bigwedge_{i=1}^d (\pi_i(j)v_0 + (1 - \pi_i(j))v_i)$$

where each of the terms on the right-hand side corresponds to the set $H_i(\pi_i(j))$ of $P \in D$ whose i^{th} coordinate is the same as that of P_j.

Hence each of the 2^d basis vectors e_i can be written as a 'polynomial' in the $v_1, \ldots, v_d$ of 'degree' at most d, that is, as a sum of terms each having at most $d-1$ $\wedge$s. We see that R is generated by the $v_{i_1} \wedge \ldots \wedge v_{i_s}$ for $0 \leq s \leq d$, as required. $\square$

Proposition 5.5.3. *The Reed–Muller code $RM(d,r)$ with $0 < r < d$ can be expressed as the bar product $RM(d-1,r)|RM(d-1,r-1)$.*

Proof. It is clear that $RM(d-1,r-1)$ is contained in $RM(d-1,r)$, so it makes sense to take the bar product.

Set up vectors in $GF(2)^d$ so that $v_d = (0,\ldots,0|1,\ldots,1)$, with 2^{d-1} 0's followed by 2^{d-1} 1's. Let z be a word in $RM(d,r)$, a sum of wedge products $v_{i_1} \wedge \ldots v_{i_s}$, some of which involve v_d. Write the sum of these terms as $y \wedge v_d$

and the sum of the remaining terms in z as x, so that x and y do not involve v_d. We have $z = x + y \wedge v_d$ with $x \in RM(d, r)$ and $y \in RM(d, r - 1)$. Let x' be the vector containing the first 2^{d-1} components of x and y' similarly the vector containing the first 2^{d-1} components of y, so that $x = (x'|x')$ and $y = (y'|y')$. We have $x' \in RM(d - 1, r)$ and $y' \in RM(d - 1, r - 1)$. Now $y \wedge v_d = (y'|0)$ and $z = (x'|x' + y')$. $\qquad\square$

Proposition 5.5.4. *The minimum weight of an RM code of order r and length 2^d is 2^{d-r}.*

Proof. The weight $w(v_1 \wedge \ldots \wedge v_r)$ is 2^{d-r} so the minimum weight is at most this.

When $r = d$, the RM code is the trivial code with minimum weight $1 = 2^0$ and when $r = 0$ the RM code is the repetition code with minimum weight $n = 2^d$.

For other values of r we use induction on d. The case $d = 1$ is trivial.

Suppose now $d > 1$ and the result is true for RM codes of length less than 2^d. By Proposition 5.5.3 we can view $RM(d, r)$ as the bar product $RM(d - 1, r)|RM(d - 1, r - 1)$. But by induction the minimum weight of $RM(d - 1, r)$ is 2^{d-1-r} and the minimum weight of $RM(d - 1, r - 1)$ is $2^{(d-1)-(r-1)} = 2^{d-r}$. Applying Proposition 5.3.2, the minimum weight of $RM(d, r)$ is $\min\{2.2^{d-1-r}, 2^{d-r}\} = 2^{d-r}$ as required. $\qquad\square$

The Reed–Muller code $RM(5, 1)$ was used by NASA for the Mariner missions to Mars: see NASA (1967).

Decoding Reed–Muller codes

Since Reed–Muller codes are bar products, you could use the method described after Proposition 5.3.2 to construct a decoder recursively. However, there is a more efficient method using successive majority verdicts.

We first show how to correct the errors in positions involving the terms of greatest degree by producing a set of orthogonal check vectors for these positions.

Let $I = \{i_1, \ldots, i_r\}$ be a subset of $\{1, \ldots, d\}$ and $v = v_{i_1} \wedge \ldots \wedge v_{i_r}$ a generator of $RM(d, r)$. Our aim is to find a set of orthogonal check vectors for v. Let $J = \{j_1, \ldots, j_{d-r}\}$ be the complement of I in $\{1, \ldots, d\}$ and put $w = v_{j_1} \wedge \ldots \wedge v_{j_{d-r}}$. The subset of D corresponding to v is the subspace $V = H_{i_1} \cap \ldots \cap H_{i_r}$ and the subset of D corresponding to w is the subspace $W = H_{j_1} \cap \ldots \cap H_{j_{d-r}}$. Clearly $D = V \oplus W$. We claim that the set of row vectors corresponding to the cosets $W + p$ of W in D gives the desired set of orthogonal check vectors for v.

First we show that if w_p is the element of R corresponding to the coset $W + p$, for any p in D, then the scalar product $(v, w_p) \equiv w(v \wedge w_p) \equiv 1$. In fact we claim that $w(v \wedge w_p) = 1$, that is, $V \cap (W + p)$ has exactly one element. Since $D = V \oplus W$, we can express p uniquely in the form $p = q + r$ with $q \in V$ and $r \in W$. Now $W + p = W + q$ so $q \in V \cap (W + p)$. Further, if $q' \in V \cap (W + p)$ then $q - q' \in W \cap V = \{0\}$, so that $q = q'$. We conclude that q is the unique element of $V \cap (W + p)$, as claimed.

Next we show that if $u = v_{k_1} \wedge \ldots \wedge v_{k_r}$ is any other generator of $RM(d, r)$ of degree r then $(u, w_p) = 0$. Let U be the subspace of D corresponding to u and put $K = \{k_1, \ldots, k_r\}$. The weight $w(u \wedge w_p)$ is equal to the number of elements in $U \cap (W + p)$. If U does not meet $W + p$ then this number is zero. Otherwise, suppose that $x \in U \cap (W + p)$. Then $W + p = W + x$, and $U \cap (W + x) = ((U - x) \cap W) + x = (U \cap W) + x$, so $U \cap (W + p)$ is a translate of the subspace $U \cap W$. But $U \cap W$ corresponds to $u \wedge w$ and this has weight 2^{d-l} where l is the number of distinct elements in the index set $J \cup K$. So $d - l \geq 1$ unless J and K are complementary subsets of $\{1, \ldots, d\}$, that is, unless $u = v$. We conclude that there is at most one value of p for which $(u, w_p) = 1$, as claimed.

We have now shown that the set of cosets $W + p$ of W in D gives a set of orthogonal check vectors for the basis element v. The number of such cosets is 2^{d-r} which is the same as the minimum distance of the code. Hence we can correct up to $\lfloor \frac{1}{2}(2^{d-1} - 1) \rfloor$ errors in the positions in the received word corresponding to the terms of degree r by use of majority verdicts.

Now subtract off the corrected terms of weight r. The resulting word can be considered as having been received from an $RM(d, r - 1)$ code. We can now repeat the process on the terms of degree $r - 1$ and so eventually correct all the errors.

Example 5.5.5. Decoding $RM(3, 1)$. The $RM(3, r)$ codes were described in Example 5.5.1. The code $RM(3, 1)$ has rank 4 and is generated by v_0, v_1, v_2 and v_3. The minimum distance is 4 and we want to be able to correct a single error.

The first stage corrects an error in the received word in one of the positions corresponding to a term of greatest degree: for example, v_1, that is, $(1, 1, 1, 1, 0, 0, 0, 0)$. In the notation of the previous paragraphs, $v = v_1$ and $w = v_2 \wedge v_3$, so $V = H_1$ and $W = H_2 \cap H_3$. The subspace W consists of the vectors 000 and 100 in $D = \mathrm{GF}(2)^3$ and there are 4 cosets $W + p$, given by

$p = 000, 001, 010$ and 011, say. The check vectors give equations

$$p = 000 \quad : \quad x_0 + x_4 = 0$$
$$p = 001 \quad : \quad x_1 + x_5 = 0$$
$$p = 010 \quad : \quad x_2 + x_6 = 0$$
$$p = 011 \quad : \quad x_3 + x_7 = 0$$

and there are 4 votes. If an error has occurred then the majority verdict will correct it.

The second and last stage corrects an error in the single term of degree zero, namely v_0. We have $w = v_1 \wedge v_2 \wedge v_3$, and $W = H_1 \cap H_2 \cap H_3$ is the zero subspace. There are 8 cosets, corresponding to p being any vector in D and the check vectors all give equations of the form $x_i = 0$. Taking the majority verdict allows the correction of up to two errors in this stage.

The Reed–Muller codes do not have good asymptotic behaviour. In Figure 5.5.2 we plot the error-control rate δ against the information rate ρ for the $RM(d, r)$ codes with $3 \leq d \leq 10$, on the same diagram (Figure 5.4.1) as the asymptotic bounds for binary codes established in §5.4. The error-control rate is $\delta = 2^{-r}$, so prescribing δ is equivalent to fixing r, and then the information rate of $RM(d, r)$ tends to 0 with increasing d (the vertical columns in the diagram). We leave it to you (Exercise 5.5.7) to show that a sequence of Reed–Muller codes with fixed information rate ρ must have δ tending to 0 with increasing d.

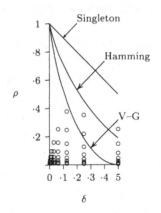

Asymptotic behaviour of Reed–Muller codes

Figure 5.5.2

Exercises

1.   Show that $RM(d, 0)$ is a repetition code.

2.   Why is the data point $(0.5, 0.25)$ in Figure 5.5.2, which represents the $RM(3, 1)$ code, above the line marked 'Hamming'?

3.   List the elements of $RM(3, 1)$. What is the relation between this code and the Hamming $[7, 16]$ code?

4.   Show that the weight enumerator of $RM(d, 1)$ is
$$y^{2^d} + \left(2^d - 2\right) x^{2^{d-1}} y^{2^{d-1}} + x^{2^d}.$$

5.   Show that $RM(d, d - 2)$ is the parity check extension of the binary Hamming $(2^d - 1, 2^d - d - 1)$ code.

6.   The dual code to $RM(d, r)$ is $RM(d, d - r - 1)$.

7.   Use Proposition 5.4.5 to show that a sequence of Reed–Muller codes with given information rate must have error-control rates tending to 0 with increasing d.

5.6   Further topics

Some historical comments on the origins of error-correcting codes are given by Thompson (1983).

We have been considering only block codes, error-control codes of fixed length. (Of course, you saw variable-length codes in Chapter 1.) An important class of variable-length codes are the 'convolutional codes', for which the encoder has a number of stages and the message affects a fixed number of the transmitted codewords. They are discussed in van Lint (1982) Chapter 11 and in McEliece (1984) Chapter II.9.

The Hamming bound 5.4.1 can be interpreted as giving the density of a 'sphere packing' defined by a code in F^n. Similarly, the Gilbert–Varshamov bound 5.4.3 gives the density of a 'covering' of F^n. From this point of view, the Second Coding Theorem (§4.2) states that a random packing is nearly as good as possible. There is a vast literature on packing and covering results, both in Euclidean space and in discrete space. You might like to browse through Conway and Sloane (1988) for an overview of some of the recent results in this area and a compendious bibliography.

Coffey and Goodman (1990) have shown that a linear code for which there is no concise specification (in the sense of algorithmic complexity) behaves like a random code.

Codes are closely related to other combinatorial structures such as graphs, block designs and association schemes. A *balanced incomplete block design (BIBD)* is a collection of subsets ('blocks') of a set of points, such that each block has the same number of points and every pair of points occur together in the same number of blocks. This occurs, for example, in the generalisation of majority-logic decoding. If we replace the requirement that at most one vector in an orthogonal check set pass an error by the requirement that at most λ pass an error then a check set of size r can correct a position if the total number of errors is at most $t = \lfloor \frac{r+\lambda-1}{2\lambda} \rfloor$; a position is in error if the number of parity checks which do not yield zero exceeds the threshold $r - \lambda(t-1)$. Cameron and van Lint (1980) discuss these relationships.

The permutation groups which act on codes are of considerable interest as many of the sporadic simple groups which arise in the classification of finite groups arise as permutation groups of codes. Thompson (1983) gives an introduction to this area and the ATLAS of Conway, Curtis, Norton, Parker and Wilson (1985) gives details.

A remarkable family of codes which are based on families of polynomials and rational functions are described by Goppa (1988) and Moreno (1991). These codes have better asymptotic behaviour than any that we are able to treat in this book. We shall give a brief description of these codes at the end of the next chapter. Their asymptotic properties are given by the following result.

Theorem 5.6.1. *Let* $q = p^{2f} \geq 49$. *There is a family of Goppa codes which exceed the asymptotic Gilbert–Varshamov bound for rates in the interval from* δ_1 *to* δ_2 *where the* δ_i *are the distinct real roots of*

$$H_q(\delta) - \delta = \frac{1}{\sqrt{q} - 1}.$$

5.7 Problems

1. Show that the information rate of an $[n, 16]$ code is at most $4/7$.

2. Show that there is no perfect 2-error correcting code of length 90 and size 2^{78} over $\{0, 1\}$.
    Hint: Assume that the zero vector is a codeword. Consider the 88 words of weight 3 with 1 in the first two places, and show that each must be distant at most 2 from a codeword.

3. Let C be a binary (n, r) code with weight enumerator $A(C, x, y)$. Show that the weight enumerator of the dual code $C^{\perp}$ is

    $$A(C^{\perp}, x, y) = 2^{-r} A(C, y - x, y + x)$$

    by considering the function

    $$g(u) = \sum_{v \in F^n} (-1)^{(u,v)} (x/y)^{w(v)}$$

    and averaging g over C.

Deduce that if C is error-correcting then the words of $C^\perp$ have average weight $n/2$. (Cambridge, 1989)

4.   Show that there is no linear $(n,3)$ code with 5 elements of weight 2 and 2 elements of weight 4.

5.   For a general $(n, n-d)$ Hamming code, where $n = 2^d - 1$, used with a binary symmetric channel with error probability p, find the weight enumerator and show that the probability of an undetected error is $(1 - 2p)^{2^{d-1}-1} - (1-p)^{2^d-1}$. How does this behave as d increases?

6.   Show that the Reed–Muller code of order $d - 2$ and length 2^d is the extension of the $(2^d - 1, 2^d - d - 1)$ Hamming code by an overall parity check digit.

7.   An *erasure* is a digit that has been made unreadable. Why are erasures easier to deal with than errors? Find necessary and sufficient conditions on the parity check matrix for a linear code that can correct t erasures, and relate t to n and k.

8.   A $(5,3)$ code over $GF(4) = \{0, 1, \alpha, \alpha^2 = 1 + \alpha\}$ has generator matrix
$$\begin{pmatrix} 1 & 0 & 0 & 1 & 1 \\ 0 & 1 & 0 & 1 & \alpha \\ 0 & 0 & 1 & 1 & \alpha^2 \end{pmatrix}.$$
Find a parity check matrix, prove that the code corrects single errors or double erasures, and show the code is perfect.

9.   Let G be a generator matrix for a binary (n, r, d) code, so that G has r rows and n columns, and suppose that exactly d of the entries in the first row are non-zero. Let G_1 be the $(r-1) \times (n-d)$ matrix formed by deleting the first row of G and those columns of G with a non-zero entry in the first row. Show that C_1, the code generated by G_1, has rank $r - 1$ and minimum distance d_1 at least $\lceil d/2 \rceil$. Deduce that
$$n \geq \sum_{i=0}^{r-1} \left\lceil \frac{d}{2^i} \right\rceil.$$
(Cambridge, 1990)

10.   Show that there is no linear ternary $(5,3)$ code with minimum weight 3.

11.   If C is a binary e-error correcting code of length n, show that C

detects $2e$ errors and show that C can be extended to a code of length $n + 1$ which detects $2e + 1$ errors.

12. Prove that a binary 2-error correcting code of length 10 can have at most 12 codewords.

13. (The Paley 2-design).

Let q be a prime congruent to 3 modulo 4 and let Q be the set of squares (quadratic residues) mod q, including 0, so that $\#Q = \frac{q+1}{2}$. Show that Q and $Q + 1$ have exactly $\frac{q+1}{4}$ elements in common and deduce that any pair of elements mod q have just $\frac{q+1}{4}$ cosets $Q + a$ in common.

Consider the code of length $q - 1$ and size $q + 1$ whose a^{th} element is $(x_1 \ldots x_{q-1})$ where $x_r = 0$ iff $r \in Q + a$, $(a = 0, \ldots, q - 1)$, and whose $(q + 1)^{\text{th}}$ element is $(1, 1, \ldots, 1)$. Show that any two distinct elements of this code are distance 5 apart.

Deduce the existence of a $[10,12]$ 2-error correcting code: cf. Problem 12 and Exercise 5.1.7.

(The codes derived in this way are not linear.)

14. (The ternary Golay code).

Let S_5 be the matrix over $GF(3)$ with first row $(0\ 1\ -1\ -1\ 1)$ and subsequent rows cyclic shifts of the first. Let $j = (11111)$; let G be

$$\begin{pmatrix} I_6 & \begin{matrix} j \\ S_5 \end{matrix} \end{pmatrix}$$

and let H be

$$\begin{pmatrix} j^{\top} & S_5 & I_5 \end{pmatrix}.$$

Show that G is the generator matrix of an $(11,6)$ code C over $GF(3)$ and H is the corresponding parity check.

No word in G has weight $\equiv 1 \bmod 3$; every non-zero word has weight ≥ 5.

Deduce that C is a perfect 2-error-correcting code.

CYCLIC CODES

6.1 Cyclic codes

Let F be a finite field $GF(q)$. We shall consider only linear codes in this chapter.

Definition. *A linear code C in F^n is cyclic if, whenever $(a_0, \ldots, a_{n-1})$ is a codeword in C, then the cyclic shift $(a_{n-1}, a_0, \ldots, a_{n-2})$ is also a codeword.*

The repetition, paper tape and Hamming $(7,4)$ codes are all examples of cyclic codes (see Exercise 5.1.3).

Identify the vector space F^n with the quotient ring $R = F[X]/\langle X^n - 1 \rangle$ by letting the vector $(a_0, \ldots, a_{n-1})$ correspond to the polynomial $a(X) = \sum_0^{n-1} a_i X^i$. Cyclic shift is represented by multiplication by X modulo $X^n - 1$. If C is a cyclic code in F^n, then $a(X) \in C$ implies $Xa(X) \in C$, so C must be an ideal of R; conversely any ideal of R corresponds to a cyclic code.

Since R is a quotient of the principal ideal domain $F[X]$ it is also a principal ideal ring. The ideal C is therefore principal: we call its generator $g(X)$ the *generator polynomial*, which must be a factor of $X^n - 1$.

We shall consider only *separable* cyclic codes, i.e. where q and n are coprime. In this case the polynomial $X^n - 1$ has a decomposition into distinct irreducible factors,

$$X^n - 1 = \prod_1^t f_i(X)$$

and so the factor g of $X^n - 1$ corresponds to one of the 2^t subsets of the f_i.

Definition. *The codes with generators f_i are maximal cyclic codes of length n, those with generator $(X^n - 1)/f_i$ are minimal cyclic.*

If $g_1(X)$ divides $g_2(X)$ then the corresponding code C_1 contains C_2. The terms 'maximal' and 'minimal' refer to this inclusion.

If $g(X) = (g_0, \ldots, g_k)$ is a generator of C, of degree k, then a basis for C is given by $g, Xg, \ldots, X^{n-k-1}g$, so C has rank $r = n - k$. If we let $a_0, \ldots, a_{k-1}$ be a message vector, the corresponding codeword is $a(X)g(X) \bmod X^n - 1$. A generator matrix for C is given by the $(n - k) \times k$ matrix

$$G = \begin{pmatrix} g_0 & g_1 & \cdots & g_k & 0 & 0 & \cdots & 0 \\ 0 & g_0 & \cdots & g_{k-1} & g_k & 0 & \cdots & 0 \\ 0 & 0 & \cdots & & & & \cdots & 0 \\ 0 & 0 & \cdots & 0 & g_0 & g_1 & \cdots & g_k \end{pmatrix}.$$

Define polynomials $r_i(X)$ by $X^i \equiv r_i(X) \bmod g(X)$. We have $r_i(X) = X^i$ for $i < k$. The polynomials $X^i - r_i(X)$ are all in C and linearly independent, hence generate C, and a generator matrix for C is $(R|I_r)$ where R is the r by k matrix with the $-r_i$ as rows for $i = k, \ldots, n - 1$. Since C is cyclic this gives $(I|R)$ as the standard form.

If $g(X)h(X) = X^n - 1$ then $g_0 h_i + g_1 h_{i-1} + \cdots + g_{n-k}h_{i-n+k} = 0$ for $i = 0, \ldots, n - k$, that is

$$H = \begin{pmatrix} 0 & 0 & \cdots & 0 & h_k & \cdots & h_1 & h_0 \\ 0 & 0 & \cdots & h_k & h_{k-1} & \cdots & 0 & 0 \\ & & \cdots & & & \cdots & & \\ h_k & h_{k-1} & \cdots & & & \cdots & 0 & 0 & 0 \end{pmatrix}$$

is a parity check matrix for the code. So the cyclic code generated by $h(X)$ is the dual code $C^\perp$ with co-ordinates reversed. Minimal and maximal codes are dual in this sense.

Let $X^n - 1 = f_1(X) \ldots f_t(X)$ and let β_i be a root of f_i in some extension of $GF(q)$. Then $f_i(X)$ is the minimal polynomial of β_i and the corresponding maximal code is just the set of polynomials $c(X)$ such that $c(\beta_i) = 0$.

Consider a set of elements $\alpha_1, \ldots, \alpha_s$ of some extension of $GF(q)$ and let $g(X)$ be the least common multiple of their minimal polynomials over $GF(q)$. Then the cyclic code with generator g consists of the polynomials $c(X)$ such that $c(\alpha_i) = 0$, $i = 1, \ldots, s$. Hence prescribing a cyclic code is equivalent to prescribing the common roots of its code words.

Definition. *A defining set for a cyclic code C over F is a set A of elements of some field K containing F such that $a(X)$ is in C iff every element of A is a root of a.*

Equivalently, C is generated by the least common multiple of the minimal polynomials of the elements of A.

Let C be a cyclic code of length n over F corresponding to a generator polynomial f. If α is a root of f, lying in some extension $\mathrm{GF}(q^d)$, then we can regard $(1, \alpha, \alpha^2 \ldots \alpha^{n-1})^\top$ formally as a parity check vector for C.

Now $\mathrm{GF}(q^d)$ is a vector space of dimension d over $\mathrm{GF}(q)$ with basis $1, \alpha, \ldots, \alpha^{d-1}$. So the 'check vector'

$$(1, \alpha, \ldots, \alpha^{n-1})^\top$$

corresponds to the matrix

$$\left(\begin{array}{cccc|c}
1 & 0 & & 0 & \\
0 & 1 & & 0 & \\
0 & 0 & \ldots & 0 & H_1 \\
0 & 0 & & 1 &
\end{array} \right)^\top$$

where H_1 has $\alpha^d \ldots \alpha^{n-1}$ as columns.

Example 6.1.1. The binary Golay code. Let $q = 2$, $n = 23$. We have the factorisation

$$X^{23} - 1 = (X + 1)(X^{11} + X^9 + X^7 + X^6 + X^5 + X + 1)$$
$$(X^{11} + X^{10} + X^6 + X^5 + X^4 + X^2 + 1)$$

into irreducible polynomials.

We note the first factor of degree 11 is

$$g_0(X) = \prod_{R_0}(X - \alpha^r)$$

where $R_0 = \{m^2 \bmod 23 \ : \ m \neq 0\}$ is the set of non-zero quadratic residues mod 23 and α is an element of $\mathrm{GF}(2^{11})$ of order 23.

The code with generator $g_0(X)$ is a $(23, 12)$ code; it can be shown to have minimum distance 7. (You will show in Problem 3 that it has minimum distance at least 5.) It is therefore 3-error-correcting and so perfect.

Exercises

1.   Identify the cyclic codes of length n corresponding to the polynomials 1, $X - 1$, $X^{n-1} + X^{n-2} + \ldots + X + 1$.

2.   Factorise $X^7 - 1$ over the field $\mathrm{GF}(2)$ and hence list all the binary cyclic codes of length 7. Identify the Hamming $(7, 4)$ code in your list.

3.   The binary Hamming code $(n, n - d)$ with $n = 2^d - 1$ is the cyclic code corresponding to a primitive n^{th} root of unity.

4.   Suppose that C is a cyclic code of length n with generator poly-
        nomial $g(X)$, and that $X - 1$ does not divide $g(X)$. Show that the
        cyclic code with generator $(X - 1)g(X)$ is the extension C^+ of C by
        a parity check digit.

6.2 BCH codes

Definition. *Let β be a primitive n^{th} root of unity in $GF(q^r)$, for some
$r \geq 1$, let $\delta \geq 2$ and let $g(X)$ be the least common multiple of the minimal
polynomials of $\beta^l, \beta^{l+1}, \ldots, \beta^{l+\delta-2}$ for some l. We see that $g(X)$ divides
$X^n - 1$. The cyclic code of length n with defining set $\{\beta^l, \beta^{l+1}, \ldots, \beta^{l+\delta-2}\}$
for some l is a* BCH *(Bose – Ray-Chaudhuri – Hocquenghem) code of* design
distance δ. *If $n = q^r - 1$ then the code is* primitive.

Equivalently, let $g(X)$ be the least common multiple of the minimal poly-
nomials of $\beta^l, \beta^{l+1}, \ldots, \beta^{l+\delta-2}$. Then g is a generator polynomial for the
BCH code. The rank of the code will depend on the degree of g.

Example 6.2.1. Let β be a primitive 7th root of unity in $GF(2^3)$, satisfying
the polynomial $X^3 + X + 1$ over $GF(2)$. Let $\delta = 2$ and $l = 1$. Consider the
BCH code with defining set $\{\beta, \beta^2\}$. Both β and β^2 are roots of the same
polynomial $X^3 + X + 1$, which generates a cyclic code of length 7 and rank
4. You will recognise it from Exercise 6.1.2 as the Hamming $(7, 4)$ code.

We must now justify the phrase 'design distance' in the definition of BCH
codes.

Lemma 6.2.2. *The determinant*

$$\begin{vmatrix} 1 & 1 & \cdots & 1 \\ x_1 & x_2 & \cdots & x_n \\ x_1^2 & x_2^2 & \cdots & x_n^2 \\ \cdots & \cdots & & \cdots \\ x_1^{n-1} & x_2^{n-1} & \cdots & x_n^{n-1} \end{vmatrix}$$

is equal to

$$\prod_{i>j}(x_i - x_j).$$

Proof. Regard the entries x_i in the determinant as independent variables.
The determinant vanishes if any of the x_i are equal, and so is divisible by the
product of all the distinct $x_i - x_j$. But the total degree of the determinant

in all the x_i is $\frac{1}{2}n(n-1)$ and so differs from the product only by a scalar factor. Equating the coefficients of the term $x_2 x_3^2 \cdots x_n^{n-1}$ coming from the main diagonal shows that this scalar is 1. $\qquad\square$

This is a *van der Monde* determinant.

Theorem 6.2.3 The BCH bound. *The minimum distance between code-words in a BCH code of design distance δ is indeed $\geq \delta$.*

Proof. Form the $n \times (\delta - 1)$ parity check matrix

$$H = \begin{pmatrix} 1 & 1 & \cdots & 1 \\ \beta^l & \beta^{l+1} & \cdots & \beta^{l+\delta-2} \\ \beta^{2l} & \beta^{2(l+1)} & \cdots & \beta^{2(l+\delta-2)} \\ \cdots & \cdots & & \cdots & \cdots \\ \beta^{(n-1)l} & \beta^{(n-1)(l+1)} & \cdots & \beta^{(n-1)(l+\delta-2)} \end{pmatrix}$$

and expand it to a $n \times r(\delta - 1)$ parity check matrix H_1 by regarding each element of $\mathrm{GF}(q^r)$ as a column vector of size r over $\mathrm{GF}(q)$. Any $\delta - 1 \times \delta - 1$ submatrix of H can be expanded by the Lemma to a power of β times a product of terms of the form $\beta^i - \beta^j$ where i and j differ by at most $\delta - 1$. But then all the $\beta^i - \beta^j$ are non-zero, and so such a submatrix is non-singular. We conclude that any $\delta - 1$ rows of H, and hence of H_1, are linearly independent over $\mathrm{GF}(q)$. But then any non-zero codeword, corresponding to a linear relation among the rows of H_1, must be of weight at least δ. $\qquad\square$

Example 6.2.4. Let $r = 5$, $q = 2$, so that $n = 31$. Set $\delta = 8$. Let α be a primitive element in $\mathrm{GF}(2^5)$. The minimal polynomial of α is $g(X) = (X - \alpha)(X - \alpha^2)(X - \alpha^4)(X - \alpha^8)(X - \alpha^{16})$, which is also the minimal polynomial of α^2, α^4 and α^8. Similarly the minimal polynomial of α^3 is also that of α^6 and the minimal polynomial of α^5 is also that of α^{10}, α^{20} and $\alpha^{40} = \alpha^9$. So the generator polynomial for C is also divisible by the minimal polynomials of α^8, α^9 and α^{10} and the code C is the same as that of design distance $\delta = 11$. Hence $d(C) \geq 11$.

Unfortunately long BCH codes are bad: if C_i is a sequence of BCH codes with parameters (n_i, k_i, d_i), then it can be shown that either $k_i/n_i \to 0$ or $d_i/n_i \to 0$.

Definition. *A* Reed–Solomon *code of length $n = q - 1$ over $F = \mathrm{GF}(q)$ is a BCH code corresponding to a primitive n^{th} root of unity in F: that is, a BCH code with $r = 1$.*

The rank of a Reed–Solomon code is $n - \delta + 1$.

Example 6.2.5. Compact disc players use two 'interleaved' Reed–Solomon codes over $\text{GF}(2^8)$. The first has length 32, the second length 28. In each case $\delta = 5$. The specification of the code calls for it to be able to completely correct a burst of about 4000 consecutive errors, about 2.5 mm of track on the disc. An error burst about three times as long can be corrected by linear interpolation of the audio signal. Further details are given by Hoeve, Timmermans and Vries (1982).

Exercises

1.   The Hamming binary $(7, 4)$ code is a BCH code. What values of n, r and δ does it correspond to?

2.   Construct a ternary BCH code of design distance 5.

3.   Show that the generator of a Reed–Solomon code has degree δ and verify that the rank is $n - \delta + 1$.

4.   Show that the information rate of the code used in the compact disc is $3/4$.

6.3   Decoding BCH codes

Let C be a BCH code of length n over $\text{GF}(q)$ with design distance $\delta = 2t + 1$. Let β be an nth root of unity in $\text{GF}(q^r)$. Suppose that a codeword c is transmitted incorrectly and received as r, with and that $r - c = e = (e_0, \ldots, e_{n-1})$ is the error vector; we interpret e as a polynomial $e(x)$ of degree at most $n - 1$. Correcting the error is equivalent to finding e.

Let P be the set of positions where an error has occurred, and p the number of errors. We assume $p \leq t$.

Definition. *The* error-locator polynomial *is the polynomial whose roots are β^{-i} for $i \in P$,*

$$\eta(x) = \prod_{i \in P} \left(1 - \beta^i x \right).$$

Finding η enables us to find P and hence to locate the errors. We further define

$$\phi(x) = \sum_{i \in P} e_i \beta^i x \, \frac{\eta(x)}{(1 - \beta^i x)}.$$

Finding both η and ϕ enables us to find the e_i and hence to correct the errors. We have

$$\frac{\eta(x)}{\phi(x)} = \sum_{i \in P} \frac{e_i \beta^i x}{(1 - \beta^i x)}$$

and the right-hand side is a partial fraction expansion of the rational function $\eta(x)/\phi(x)$.

If this rational function were defined over the real or complex numbers then we would be justified in taking a power series expansion. Over the finite field $\mathrm{GF}(q^r)$ we simply proceed formally. We write

$$\frac{1}{(1 - \beta^i x)} = \sum_{j=1}^{\infty} \left(\beta^i x\right)^j,$$

so that

$$\frac{\eta(x)}{\phi(x)} = \sum_{i \in P} \sum_{j=1}^{\infty} e_i \left(\beta^i x\right)^j$$

$$= \sum_{j=1}^{\infty} x^j \sum_{i \in P} e_i \left(\beta^{ij} x\right)$$

$$= \sum_{j=1}^{\infty} x^j e \left(\beta^j x\right)$$

By definition of the BCH code, the codeword $c(x)$ has β^j as roots for $j = 1, \ldots, 2t$. So for these values of j we have $e\left(\beta^j x\right) = r\left(\beta^j x\right)$ and $r(x)$ is known. Put $R_j = r\left(\beta^j x\right)$ and let $\eta(x) = \eta_0 + \ldots + \eta_p x^p$. We have

$$\phi(x) = \sum_{j=1}^{2t} R_j x^j \quad \sum_{i=0}^{p} \eta_i x^i$$

at least for powers up to x^{2t} and deduce that

$$\sum_{i+j=k} S_j \eta_i = 0 \qquad \text{for } p + 1 \leq j \leq 2t$$

which enables us to find the η_i by solving a set of simultaneous linear equations. Finally ϕ is recovered by substituting for η in the previous equation.

Exercise

1.   Use Lemma 6.2.2 to show that the system of linear equations giving the error-locator polynomial is non-singular.

6.4   Further topics

There are stronger results on the minimum distance of a cyclic code than the BCH bound. Van Lint and Wilson (1986) have given a summary and tabulate those codes of small length for which the BCH bound is not sharp. They establish the following result, which may help you with Problem 6.5.4. Let A be a defining set for a code of minimum distance d and let $B = \{\beta^{i_1}, \ldots, \beta^{i_j}\}$ be a collection of powers of an nth root of unity β, with the exponents i in increasing order. Let $k = (i_j - i_1) - (j - 1)$ be the number of powers of β 'missed out' in B. If $k < d - 1$, then the set of products AB is a defining set for a code of minimum distance at least $d + j - 1$.

The binary Golay code, Example 6.1.1, is a *quadratic residue code*, where the defining set is of the form $\{\alpha^{i^2}\}$. Such codes have surprisingly large symmetry groups. That of the parity digit extension of the binary Golay code is the Mathieu group M_{24} of order 244823040. The ATLAS of Conway, Curtis, Norton, Parker and Wilson (1985) gives constructions for many such groups.

We now give a few details of the Goppa codes mentioned in the last chapter. Let F be the finite field $\mathrm{GF}(q)$ and K the finite field $\mathrm{GF}(q^m)$. Let $g(X)$ be a polynomial of degree t over F and $A = \{\alpha_1, \ldots, \alpha_n\}$ a set of n distinct elements of K, not roots of g. The *Goppa code* for g and A is the set of vectors $c \in F^n$ such that

$$\sum_1^n \frac{c_i}{X - \alpha_i} \equiv 0 \bmod g(X).$$

This code has weight at least $t + 1$ and rank at least $n - mt$. It may be decoded by a construction similar to that of §3.

We may view this construction as defining a vector space of rational functions with prescribed 'poles' at the points α_i. Goppa's generalisation was to consider spaces of functions defined on algebraic curves and with prescribed poles. From this point of view, BCH codes come from functions on the curve whose general point is of the form $(1, x, x^2, \ldots, x^{6-1})$. The Hamming $(7, 4)$ code can be obtained as a space of functions on the curve

$$X + X^3Y + Y^3 + X^2 + X^2Y^2 + Y^2 + XY + X^2Y = 0$$

over $\mathrm{GF}(2)$ having poles at the seven (projective) points on this curve.

6.5 Problems

1.   Show that over $\mathrm{GF}(2)$ the $(7,4,3)$ code with $g_1(X) = X^3 + X + 1$ is dual to the $(7,3,4)$ code with $g_2(X) = X^4 + X^3 + X^2 + 1$. Write down all the codewords of the latter code.

2.   Let α be a primitive $(2^m - 1)^{\text{th}}$ root of unity, with minimal polyno-
        mial $M(X)$. Use the BCH bound to show that a $(2^m - 1, m)$ cyclic
        binary code with parity check polynomial $M(X)$ has minimum dis-
        tance at least 2^{m-1}.

3.   Why might one suspect the existence of a perfect binary $(23, 12, 7)$
        code?
            Consider the generator polynomial
$$g(X) = (X - \beta)(X - \beta^2)(X - \beta^4)\dots(X - \beta^{2^{10}}),$$
        where β is an element of GF(2048) having order 23; on multiplying
        out the brackets (in GF(2048)) g has coefficients in GF(2). Show
        that the code generated by this polynomial has minimum weight at
        least 5.

4.   Let α be a root of unity of order 21 in GF(64) and let C be the
        binary code of length 21 with defining set $\{\alpha, \alpha^3, \alpha^7, \alpha^9\}$. Show that
        C has minimum distance 8.

APPENDIX: RINGS, FIELDS AND VECTOR SPACES

This appendix is intended you remind you of the algebraic objects you will be meeting, especially in Chapters 5 and 6. We have attempted to give definitions of all the algebraic objects you are likely to meet in the main part of the book, mainly in order to fix our terminology and notation. We certainly do not expect you to use it as your first introduction to abstract algebra!

We shall take the notion of Abelian group as fundamental.

7.1 Rings

Definition. *A ring* $(R, +, -, 0, \times, 1)$ *is an Abelian group with respect to addition* $+$, *negation* $-$ *and with an additive identity* 0. *The operation* $\times$ *is associative and commutative, and* 1 *is an identity element for* $\times$. *Further,* $\times$ *is distributive over addition, that is*

$$x \times (y + z) = x \times y + x \times z$$
$$(y + z) \times x = y \times x + z \times x.$$

We require that $1 \neq 0$. *The ring is commutative if* $x \times y = y \times x$.

We almost always omit the $\times$ symbol and denote multiplication by juxtaposition.

We are only going to consider commutative rings and shall often omit the word 'commutative'.

Example 7.1.1. The set of integers $\mathbb{Z}$, with the usual operations of addition and multiplication, forms a commutative ring.

Example 7.1.2. Let m be a positive integer. The set of residue classes modulo m, with addition and multiplication taken modulo m, forms a ring $\mathbb{Z}/m$.

Example 7.1.3. The collection $\mathcal{P}X$ of all subsets of a given set X, the *power set* of X, forms a commutative ring under the operation of symmetric difference as addition and intersection as multiplication.

Definition. *The* characteristic *of a ring is* 0 *if* $1 + 1 + \ldots + 1$ *is never equal to* 0. *Otherwise, the characteristic is* n *where* n *is the least number such that*

$$\underbrace{1 + 1 + \ldots + 1}_{n \text{ times}} = 0.$$

Definition. *An* integral domain *is a commutative ring in which the product of non-zero elements is non-zero.*

The characteristic of an integral domain, if non-zero, must be a prime number.

Definitions. *An* ideal I *in a ring* R *is an additive subgroup of* R *which is closed under multiplication by any element of* R, *so that if* $x \in I$ *and* $a \in R$ *then* $ax \in I$. *A* principal ideal *with generator* x *is the ideal consisting of all multiples of* x,

$$\langle x \rangle = \{ax \ : \ a \in R\}.$$

In any ring R the sets $O = \{0\}$ and R itself are ideals. They are principal, being $\langle 0 \rangle$ and $\langle 1 \rangle$ respectively.

If S is any subset of a ring R then the set of all finite sums of the form $\sum a_i s_i$ with $a_i \in R$ and $s_i \in S$ constitutes an ideal, denoted $\langle S \rangle$.

Definition. *A* principal ideal ring *is one in which every ideal is principal. A* principal ideal domain *is an integral domain which is a principal ideal ring.*

The ring of integers $\mathbb{Z}$ is a principal ideal domain: the least positive element of an ideal can be taken as a generator.

Definition. *An element* x *of a ring* R divides *an element* y *if there is* $z \in R$ *such that* $y = xz$. *We also say that* x *is a* factor *of* y, *or that* y *is a* multiple *of* x.

Definitions. *A* unit *in a ring is an element which has a multiplicative inverse. An* irreducible *element of a ring is one which can only be expressed as a product in which one of the factors is a unit, that is, has no non-trivial factors.*

The definition of 'irreducible' applied to the ring $\mathbb{Z}$ is what you are used to calling a 'prime' number. Unfortunately in ring theory 'prime' has a different definition.

Example 7.1.4. For any ring R, we define a polynomial over R in the indeterminate X to be a finite sum of the form $f(X) = \sum a_i X^i$ with the $a_i \in R$. Making the usual definitions of addition and multiplication, the set $R[X]$ of polynomials becomes a ring. The *degree* ∂f of a non-zero polynomial f is the highest index on a non-zero coefficient (we conventionally define $\partial 0$ to be $-\infty$).

If the ring R is commutative, then so is $R[X]$. If R is an integral domain then so is $R[X]$.

Definition. *A (ring)* homomorphism *from a ring R to a ring S is a map from R to S which maps 0 to 0, 1 to 1 and which preserves sums and products. The* kernel *of a homomorphism is the set of elements mapped to 0.*

The kernel of a homomorphism is an ideal.

Definition. *The* quotient ring R/I *of a ring R by an ideal I is the set of cosets $I + x$ given a ring structure by addition $(I + x) + (I + y) = I + x + y$ and multiplication $(I + x)(I + y) = I + xy$.*

The map that sends x to $I + x$ is a ring homomorphism from R onto R/I.

Example 7.1.5. The map from $\mathbb{Z}$ to $\mathbb{Z}/m$ which sends an integer x to the residue class x mod m is a ring homomorphism with kernel $\langle m \rangle$.

The quotient ring $\mathbb{Z}/\langle m \rangle$ is just the ring $\mathbb{Z}/m$ of integers modulo m.

Exercises

1. A ring of characteristic two is necessarily commutative.

2. The image of a principal ideal ring under a ring homomorphism is again a principal ideal ring.

3.   Let p be a prime number. Show that in a ring of characteristic p we
        have $(x + y)^p = x^p + y^p$.

4.   Verify that addition and multiplication of cosets turns the quotient
        R/I into a ring.

7.2 Fields

Definition. *A field is a commutative ring in which every non-zero element
is a unit, that is, has a multiplicative inverse.*

   A field is necessarily an integral domain. The characteristic of a field is
therefore either zero or a prime number p.

Example 7.2.1. The rational numbers $\mathbb{Q}$, the real numbers $\mathbb{R}$ and the
complex numbers $\mathbb{C}$ form fields under the usual operations. In each case
the characteristic is zero.

Example 7.2.2. The integers modulo p, where p is a prime number, form
a field, $\mathbb{Z}/p$. (You will see below how to find inverses in $\mathbb{Z}/p$.) The characteristic is p.

Exercise

1.   A finite integral domain is a field.

7.3 Euclidean domains

We saw that $\mathbb{Z}$ was a principal ideal domain by consideration of the least
non-negative element in a ideal. We introduce the idea of a Euclidean
domain in order to generalise this sort of argument.

Definition. *A Euclidean domain E is an integral domain equipped with a
function $\phi : E \to \mathbb{N}$ taking non-negative integer values such that (i) for all
x, y, $\phi(xy) = \phi(x)\phi(y)$; (ii) $\phi(1) = 1$, and $\phi(x) = 0$ iff $x = 0$; (iii) for all x
and for all non-zero y there exist q and r with $x = qy + r$ and $\phi(r) < \phi(y)$.
We call ϕ a Euclidean function for E.*

Definition. *If $x = qy + r$ with $\phi(r) < \phi(y)$ then q is the* quotient *and r is
the* remainder *on division of x by y.*

Example 7.3.1. The ring of integers $\mathbb{Z}$ forms a Euclidean domain with the function $\phi(x) = |x|$. This example shows that the remainder r is not necessarily uniquely determined: for example in $\mathbb{Z}$ if $r \neq 0$ then you could also take $r - q$ as remainder.

Example 7.3.2. The ring $F[X]$ of polynomials over a field F forms a Euclidean domain with the function $\phi(f) = \exp(\partial f)$.

Every Euclidean domain is a principal ideal domain. Given an ideal I of E, an element of I with the smallest non-zero value of ϕ will be a generator of I, so I is principal.

Definition. *A highest common factor of two elements x, y of a ring R is an element $h = \mathrm{hcf}(x, y)$ such that h is a factor of both x and y and, if c is any other common factor of x, y then c divides h. Elements x, y are coprime if they have a highest common factor of 1.*

There is a corresponding definition for *least common multiple*. Not every ring has highest common factors for every pair of elements, but they always exist in a Euclidean domain.

The *Euclidean algorithm* finds the highest common factor of two elements in a Euclidean domain. Given x, y, both non-zero, define sequences q_n, r_n as follows:

$$x = yq_0 + r_1$$
$$y = r_1q_1 + r_2$$
$$\cdots$$
$$r_{n-1} = r_nq_n + r_{n+1}$$

where $\phi(r_{n+1}) < \phi(r_n)$. Since the values $\phi(r_n)$ form a decreasing sequence of positive integers, there must be an N such that $\phi(r_{N+1}) = 0$, which by condition (ii) means that $r_{N+1} = 0$. It is not hard to see that r_N is the highest common factor of x and y. Further, since each r_{n+1} is a linear combination of r_{n-1} and r_n, we can find a, b such that $r_N = ax + by$, so r_N is a generator of the ideal $\langle x, y \rangle$. We complete the algorithm in trivial cases by defining $\mathrm{hcf}(0, x) = x$.

We could find a, b from the Euclidean algorithm by storing all the values of q_n and r_n and substituting backwards. However, there is a method which does not require keeping the values. Define sequences P_n and Q_n as follows:

$P_{-1} = 1$, $Q_{-1} = 0$, $P_0 = 0$, $Q_0 = 1$ and for $n \geq 1$

$$P_{n+1} = P_{n-1} - q_n P_n$$
$$Q_{n+1} = Q_{n-1} - q_n Q_n.$$

The rule for forming successive elements is the same for the two sequences, only the initial conditions differ. If we define $x = r_{-1}$, $y = r_0$ then by induction $r_n = P_n x + Q_n y$. Hence $a = P_N$, $b = Q_N$.

If p is a prime number and x is not divisible by p, then the highest common factor of x and p is 1. Hence the Euclidean algorithm gives us a way of finding a and b such that $ax + bp = 1$. Then $a \bmod p$ is the inverse of $x \bmod p$ in the ring $\mathbb{Z}/p$. As stated above, every non-zero element of this ring has an inverse, so $\mathbb{Z}/p$ is a field.

Theorem 7.3.3. *Every element of a Euclidean domain has a factorisation into irreducible elements which is unique up to order of the factors and multiplication by units.* □

You will be familiar with this result for $\mathbb{Z}$, again with 'irreducible' replaced by 'prime'. A domain with this property is called a *unique factorisation domain*.

Exercises

1.   Let F be the field $\mathbb{Z}/2$ and let R be the Euclidean domain $F[X]$. Show that $X^2 + X + 1$ is irreducible in R.
       Factorise the polynomial $X^{23} - 1$ in R.

2.   Let x, y be elements of a Euclidean domain and h a highest common factor. Show that xy/h is a least common multiple.

3.   Suppose that x, y are coprime positive integers. Show that every integer N greater than xy can be expressed in the form $N = ax + by$ where a and b are *non-negative* integers.

7.4   Vector spaces

Definition. *A vector space V over a field F is an Abelian group with an action of F on V, scalar multiplication, denoted av for $a \in F$ and $v \in V$ satisfying (i) $a(v + w) = av + aw$; (ii) $(a + b)v = av + bv$; (iii) $(ab)v = a(bv)$; (iv) $1v = v$.*

Example 7.4.1. The collection of row (or column) vectors of length n

with entries in a field F form a vector space, denoted F^n over F with the operations

$$(x_i) + (y_i) = (x_i + y_i); \qquad a(x_i) = (ax_i).$$

In particular, F is a vector space over itself.

Definitions. *A* linear combination *of a finite subset $\{v_1, \ldots, v_n\}$ of a vector space V over a field F is an expression of the form $\sum_i a_i v_i$ with the a_i in F. A* linear relation *on a finite subset S is a relation of the form $\sum_i a_i v_i = 0$. A subset S of a vector space V is* linearly independent *if the only linear relations that exist on S have all the coefficients a_i equal to 0. A subset S* spans *V if every element of V is expressible as a linear combination of S. A* basis *or* base *for V is a linearly independent spanning set.*

Theorem 7.4.2. *If V has a finite spanning set, then it has a basis. If V has a basis, then all bases have the same number of elements.* $\square$

Definition. *The* dimension *of a vector space is the number of elements in a basis, if it exists.*

The space of row vectors F^n has dimension n. The standard basis consists of the vectors e_i with zeroes in all entries except for a 1 in the ith place.

Definition. *A* subspace *of a vector space V is an additive subgroup U which is closed under scalar multiplication.*

The zero subspace $\{0\}$ and V itself are always subspaces of V.

Definitions. *Subspaces X, Y of V are* linearly disjoint *if the intersection $X \cap Y$ is $\{0\}$. We call X, Y* complementary *if X and Y are linearly disjoint and every element of V can be written as the sum of an element of X and an element of Y. If so, V is the* (internal) direct sum *of X and Y, denoted $V = X \oplus Y$.*

If $V = X \oplus Y$ then the dimension of V is the sum of the dimensions of X and Y and the representation of each element of V as the sum of an element of X and an element of Y is unique.

Definition. *Let U be a subspace of V. The* coset *of an element x of V is*

$$U + x = \{u + x \ : \ u \in U\}.$$

The cosets $U + x$ and $U + y$ are equal if and only if $x - y \in U$. The coset $U + 0$ is U itself. The cosets of U partition V.

Definition. A linear map from a vector space V to another vector space W, both over the same field F, is a map T from V to W satisfying (i) $T(v + w) = T(v) + T(w)$; (ii) $T(av) = aT(v)$. The set of linear maps from V to W is denoted $L(V, W)$. A linear functional on a vector space V is a linear map from V to F regarded as a vector space over itself.

Linear maps form a vector space under the *pointwise* operations $(f + g)(v) = f(v) + g(v)$ and $(af)(v) = a(f(v))$.

Definition. The dual space V^* of a vector space V is the vector space $L(V, F)$ of linear functionals on V.

If V has dimension n then so does V^*.

Definition. The annihilator of a subspace U of V is the subspace of V^*

$$U^\circ = \{f \in V^* \ : \ \forall u \in U \quad f(u) = 0\}.$$

If V has dimension n and U has dimension r then U° has dimension $n - r$.

Exercises

1.   If V has dimension n and W has dimension m then $L(V, W)$ has dimension mn.

2.   Show that there is a natural correspondence between V and V^{**}.

7.5   Finite fields

Suppose that F is a finite field, and K is a finite field containing F as a subfield. Then K is a vector space over F, necessarily of finite dimension, say f. Hence if $\#F = q$ then $\#K = q^f$. We say that K is an *extension* of F of *degree* f.

Theorem 7.5.1. *There is (up to isomorphism) just one finite field of each prime power order.*   □

We call this field $\mathrm{GF}(q)$. Of course $\mathrm{GF}(p^1) = \mathbb{Z}/p$ is the ring of integers modulo the prime p. The *Frobenius* map $\phi : x \to x^p$ is an *automorphism* of

GF(q), an isomorphism of GF(q) with itself (Exercise 7.1.3). NB: the field
GF(p^f) for $f > 1$ is **not** $\mathbb{Z}/p^f$; the latter is not even an integral domain.

If you know some Galois theory, you can check that the extension GF(q) is Galois
over GF(p) with cyclic Galois group generated by ϕ.

Example 7.5.2. Let α be an element of GF(4) which is not 0 or 1. Then
$\alpha^2 \neq \alpha$, 0 or 1 and $\alpha + 1 \neq \alpha$, 0 or 1. So $\alpha^2 = \alpha + 1$ because there is only
one further element of the field.

$$\text{GF}(4) = \{0, 1, \alpha, \alpha^2 = \alpha + 1\}.$$

Let $f(X)$ be an irreducible polynomial of degree n over the field $F = $
GF(q). The quotient ring $K = \text{GF}(q)[X]/\langle f \rangle$ is a finite integral domain
and hence (Exercise 7.2.1) a field. If we let x denote the coset of X modulo
$\langle f \rangle$ then $1, x, \ldots, x^{n-1}$ are linearly independent over F and hence K is an
extension of F of degree n. So by Theorem 7.5.1, K is simply GF(q^n).

Let F be a finite field, and K a finite field containing F, of dimension
f as a vector space over F. Let β be an element of K. Then the $f + 1$
elements $1, \beta, \ldots, \beta^f$ are linearly dependent over F and hence β satisfies
some polynomial equation with coefficients in F and degree at most f. The
set of such polynomials forms an ideal in $F[X]$ and so is generated by one
of its non-zero elements of least degree, a *minimal polynomial* for β over F.

Theorem 7.5.3. *The non-zero elements of a finite field form a cyclic group
under multiplication.* □

Again we are not going to attempt to prove this result. For proofs of both the
Theorems, see Lidl and Niederreiter (1983) Chapter 2.1.

Definitions. *A* primitive element *of a finite field is a generator of the
multiplicative group. A* primitive r^{th} root of unity *is an element of a field
of order exactly r in the multiplicative group.*

Since every element of the multiplicative group of GF(q) satisfies $x^{q-1} = $
1, a primitive element is just a primitive $(q-1)^{\text{th}}$ root of unity.

Exercises

1. Show that $X^2 + X + 1$ is the only irreducible polynomial of degree
   2 over GF(2) and hence recover the description of GF(4) given in
   Example 7.5.2.

2. Factorise the polynomial $X^7 - 1$ over the field GF(2). Hence show that the field GF(8) must contain an element α such that $\alpha^3 = \alpha + 1$. Show that $\{1, \alpha, \alpha^2\}$ forms a basis for GF(8) as a vector space over GF(2) and hence construct the addition and multiplication tables.

3. Every element of GF(q) is a root of $X^q - X$.

7.6 Problems

1. Let B be a ring in which $xx = x$ for all x. Show that B has characteristic 2 and is commutative.

    Define an order relation $\leq$ on B by defining $x \leq y$ iff $xy = x$. Define an element x to be an *atom* if $0 \leq y \leq x$ implies that $y = 0$ or $y = x$.

    Show that if B is finite then B can be regarded as $\mathcal{P}X$ where X is the set of atoms of B. Give a counter-example in the case when B is infinite.

2. Show that every irreducible polynomial of degree n over GF(q) must divide $X^{q-1} - 1$. Show that there is an irreducible polynomial of every degree over GF(q) and deduce that there exists at least one finite field of every prime power order.

BIBLIOGRAPHY

Further reading

Ash, R. B. (1965): *Information Theory*. Interscience Tracts in Pure and Applied Mathematics **19**, Interscience, New York. *[Although out of print, this text is well worth tracking down in any library you have the use of. It is remarkably clear as well as precise, and can be relied upon for correct probability arguments]*

Billingsley, P. (1965): *Ergodic Theory and Information*. Wiley Ser. in Prob. & Math. Statistics, Wiley, New York. *[The classic text on ergodic theory. It develops entropy in depth, establishing its importance in general measure-preserving transformations]*

Blahut, R. E. (1987): *Principles and Practice of Information Theory*. Addison-Wesley, Reading, Mass. *[The best recent book on information theory from an engineering perspective. It discusses and makes sense of a huge range of recent work, and is strong on connections with statistical inference]*

Chambers, W. G. (1985): *Basics of Communication and Coding*. Oxford Science Publications, Clarendon Press, Oxford. *[Information theory from an applied-mathematics perspective, together with some coding theory and cryptography]*

Cover, T. M., Gacs, P. & Gray, R. M. (1989): 'Kolmogorov's contributions to information theory and algorithmic complexity'. *Ann. Probab.* **17**, 840–65. *[Worth reading for its broad, authoritative and recent viewpoint]*

Csiszár, I. & Körner, J. (1981): *Information Theory: Coding Theorems for Discrete Memoryless Systems*. Academic Press, Orlando, Fla.

[Advanced text containing much material, written in a rigorous mathematical style]

Gray, R. M. (1990a): *Source Coding Theory.* Kluwer Internat. Series in Eng. & Comp. Sci., Kluwer, Dordrecht. *[Recent treatise on an intricate part of information theory, informed by engineering considerations]*

Gray, R. M. (1990b): *Entropy and Information Theory.* Springer-Verlag, New York. *[A new advanced text]*

Hill, R. (1986): *A First Course in Coding Theory.* Oxford Appl. Mathematics & Comp. Sci. Series, Clarendon Press, Oxford. *[Extends the material on error-correcting codes and gives up-to-date bounds]*

Lidl, R. & Niederreiter, H. (1983): *Finite Fields.* Encycl. of Mathematics and its Appl. **20**, Cambridge Univ. Press, Cambridge. *[A compendious reference work for finite fields including much on linear codes]*

van Lint, J. H. (1982): *Introduction to Coding Theory.* Graduate Texts in Mathematics **86**, Springer-Verlag, New York. *[A very readable account of the algebraic side of coding theory]*

Longo, G. (ed.) (1977): *The Information Theory Approach to Communications.* Springer-Verlag, New York. *[Review articles on the state of information theory at the time of publication]*

McEliece, R. J. (1984): *The Theory of Information and Coding: a Mathematical Framework for Communication.* Encycl. of Mathematics and its Appl. **3**, Cambridge Univ. Press, Cambridge. *[Advanced text on both information theory and coding]*

MacWilliams, F. J. & Sloane, N. J. A. (1977): *The Theory of Error-correcting Codes.* North-Holland Math. Library **16**, North-Holland, Amsterdam. *[An encyclopaedic work]*

Rényi, A. (1987): *A Diary on Information Theory.* Wiley, Chichester. *[Much insight to be gained here, in an afternoon's reading]*

Shannon, C. E. & Weaver, W. (1949): *The Mathematical Theory of Communication.* Reprinted 1963, Univ. of Illinois Press, Urbana, Ill. *[A reprint of Shannon's paper of 1948, that instantly created a new mathematical subject. Very readable, and not very technical]*

Storer, J. A. (1988): *Data Compression: Methods and Theory.* Principles of Comp. Sci. Series **13**, Computer Science Press, Rockville, MD. *[Discusses implementation, Lempel–Ziv coding, image processing]*

Welsh, D. (1988): *Codes and Cryptography.* Oxford Science Publications, Clarendon Press, Oxford. *[At about the level of this book, a broad coverage emphasizing complexity and cryptography]*

References

Most of the references below will provide good further reading on specific topics. You will find following each reference a note of the section or sections where it is mentioned.

Abramowitz, M. & Stegun, I.A. (1975): *Handbook of Mathematical Functions.* Dover, New York. [§5.4]

Aigner, M. (1988): *Combinatorial Search.* Wiley-Teubner Ser. Comp. Sci., Wiley, New York. [§1.8]

Algoet, P. H. & Cover, T. M. (1988): 'A sandwich proof of the Shannon-McMillan-Breiman theorem'. *Ann. Probab.*, **16**, 899–909. [§2.12]

Blake, I. F. (1982): 'The enumeration of certain run-length sequences'. *Info. & Control*, **55**, 222–37. [§1.8]

British Standards Institution (1985): *BS 6692: Coded character sets for text communication.* British Standards Institution, London. [§5.1]

Cameron, P. J. & van Lint, J. H. (1980): *Graphs, Codes and Designs.* London Math. Soc. Lecture Notes **43**, Cambridge Univ. Press, Cambridge. [§5.6]

Campbell, L. L. (1965): 'A coding theorem and Rényi's entropy'. *Info. & Control*, **8**, 423–9. [§1.9]

Carl, B. & Stephani, I. (1990): *Entropy, Compactness and the Approximation of Functions.* Cambridge Tracts in Mathematics **98**, Cambridge Univ. Press, Cambridge. [§2.13]

Coffey, J. T. & Goodman, R. M. (1990): 'Any code of which we cannot think is good'. *IEEE Trans. Info. Th.*, **IT-36**, 1453–1461. [§5.6]

Conway, J. H., Curtis, R. T., Norton, S. P., Parker, R. A. & Wilson, R. A. (1985): ATLAS *of Finite Groups.* Clarendon Press, Oxford. [§5.6, §6.4]

Conway, J. H. & Sloane, N. J. A. (1988): *Sphere Packings, Lattices and Groups.* Grundlehren der mathematischen Wissenschaften **290**, Springer-Verlag, New York. [§5.6]

Cover, T. M. (1972): 'Broadcast channels'. *IEEE Trans. Info. Th.*, **IT-18**, 2–14. [§4.7]

Cover, T. M. & Thomas, J. A. (1988): 'Determinant inequalities via information theory'. *SIAM J. Matrix Anal. Appl.*, **9**, 384–92. [§4.7]

Davisson, L. D. (1973): 'Universal noiseless coding'. *IEEE Trans. Info. Th.*, **IT-19**, 783–95. [§1.8]

Davisson, L. D., McEliece, R. J., Pursley, M. B. & Wallace, M. S. (1981): 'Efficient universal noiseless source codes'. *IEEE Trans. Info. Th.*, **IT-27**, 269–79. [§1.8]

Dudley, R. M. (1974): 'Metric entropy of some classes of sets with differentiable boundaries'. *J. Approx. Th.*, **10**, 227–36. [§2.13]

El Gamal, A. & Cover, T. M. (1980): 'Multiple user information theory'. *Proc. IEEE*, **68**, 1466–83. [§4.7]

Feller, W. (1970) *An Introduction to Probability Theory and its Applications I*, 3rd ed., revised printing. Wiley, New York. [§5.4]

Goppa, V. D. (1988) *Geometry and Codes.* Kluwer, Dordrecht. [§5.6]

Guiaşu, S. & Shenitzer, A. (1985): 'The principle of maximum entropy'. *Math. Intelligencer*, **7**, 42–8. [§4.5]

Hamming, R. W. (1980): *Coding and Information Theory*, 2nd ed. Prentice-Hall, Englewood Cliffs, NJ. [§1.4]

Hoeve, H., Timmermans, J. & Vries, L. B. (1982): 'Error correction and concealment in the Compact Disc system'. *Philips Tech. Rev.*, **40**, 166–172. [§6.2]

Hu Kuo-Ting (1962): 'On the amount of information'. *Theor. Prob. Appl.*, **7**, 439–47. [§4.1]

Jones, C. B. (1981): 'An efficient coding system for long source sequences'. *IEEE Trans. Info. Th.*, **IT-27**, 280–91. [§1.8]

Knuth, D. E. (1973): *The Art of Computer Programming*, vol. 3, *Sorting and Searching*. Addison-Wesley, Reading, Mass. [§1.8]

Kolmogorov, A. N. & Tikhomirov, V. M. (1961): 'ε-entropy and ε-capacity of sets in a functional space'. *Amer. Math. Soc. Translations*, 2nd series, **17**, pp. 277–364. Amer. Math. Soc., Providence, R.I. [§2.13]

Kullback, S. & Leibler, R. A. (1951): 'On information and sufficiency'. *Ann. Math. Statist.*, **22**, 79–86. [§2.14]

Loève, M. (1977): *Probability Theory*, 4th ed., vol. 1. Springer-Verlag, New York. [§2.3]

Lorentz, G. G. (1966): 'Metric entropy and approximation'. *Bull. Amer. Math. Soc.*, **72**, 903–37. [§2.13]

Moreno, C. J. (1991): *Algebraic Curves over Finite Fields*. Cambridge Tracts in Mathematics **97**, Cambridge Univ. Press, Cambridge. [§5.6]

National Aeronautics and Space Administration (1967): 'Information systems'. *NASA STAR Technical Report*, **05**, 67N14438. [§5.5]

Ornstein, D. S. & Weiss, B. (1991): 'Statistical properties of chaotic systems'. *Bull. Amer. Math. Soc. (NS)*, **24**, 11–116. [§2.13]

Picard, C.-F. (1980): *Graphs and Questionnaires*. North-Holland Math. Studies **32**, North-Holland, Amsterdam. [§1.8]

Pinsker, M. S. (1964): *Information and Information Stability of Random Variables and Processes*, transl. A. Feinstein. Holden-Day, San Francisco. [§1.8]

Pólya, G. & Szegö, G. (1972): *Problems and Theorems in Analysis*, vol. I. Grundlehren der mathematischen Wissenschaften **193**, Springer-Verlag, Berlin. [§2.6]

Rényi, A. (1970): *Probability Theory*. North-Holland Ser. Appl. Mathematics & Mech. **10**, North-Holland, Amsterdam. [§1.8]

Rissanen, J. J. (1976): 'Generalized Kraft inequality and arithmetic coding'. *IBM J. Res. Develop.*, **20**, 198–203. [§1.8]

Rissanen, J. J. (1984): 'Universal coding, information, prediction, and estimation'. *IEEE Trans. Info. Th.*, **IT-30**, 629–36. [§1.8]

Robert, C. (1990): 'An entropy concentration theorem: applications in artificial intelligence and descriptive statistics'. *J. Appl. Probab.*, **27**, 303–13. [§2.13]

Schwartz, E. S. (1964): 'An optimum encoding with minimum longest code and total number of digits'. *Info. & Control*, **7**, 37–44.   [§1.9]

Stam, A. J. (1959): 'Some inequalities satisfied by the quantities of information of Fisher and Shannon'. *Info. & Control*, **2**, 101–12.   [§4.7]

Standard Book Numbering Agency (1985): *International Standard Book Numbering*. Standard Book Numbering Agency, London. [§5.1]

Thomasian, A. J. (1960): 'An elementary proof of the AEP of information theory'. *Ann. Math. Statist.*, **31**, 452–6.   [§2.12]

Thompson, T. M. (1983): *From Error-correcting Codes through Sphere Packings to Simple Groups*. Carus Math. Monographs **21**, Math. Assoc. America, Washington, D.C.   [§5.6, §5.6]

Welch, T. A. (1984): 'A technique for high-performance data compression'. *IEEE Computer*, **17.6**, 8–19.   [§1.8]

Zehavi, E. & Wolf, J. K. (1988): 'On run-length codes'. *IEEE Trans. Info. Th.*, **IT-34**, 45–54.   [§1.8]

Zubay, G. (1988): *Biochemistry*, 2nd ed. Macmillan, New York.   [§2.8]

NOTATION SUMMARY

Set notation, logic, meta-objects

$\emptyset$	empty set
$A \cup B$	union of A and B
$A \cap B$	intersection of A and B
$A \subseteq B$	A is a subset of B
A^c	complement of A
$A \setminus B$	set-theoretic difference $A \cap (B^c)$
$\# A$	number of elements in the set A
$\mathcal{P} A$	power set of A, the set of all subsets of A
$\mathbf{1}\{T\}, \mathbf{1}_T$	indicator function, 1 if T is true, 0 if not
$f : A \to B$	function from (domain) A to (range) B, $f(a) \in B$ for all $a \in A$
$a \mapsto b$	a maps to b: $f(a) = b$
$\{a : T\}$	set of all a such that T is true
$:=$	is defined to be
$f \circ g$	composition of f and g: $f \circ g(a) = f(g(a))$
$\forall$	for all

Sets of numbers

$\mathbb{N}$	the set of natural numbers 1, 2, ...
$\mathbb{Z}$	the set of integers
$\mathbb{R}$	the set of real numbers
$(x, y]$	the interval $\{t \in \mathbb{R} : x < t \leq y\}$; similarly for $[x, y)$, etc.
$\mathbb{Z}/m$	the set of integers modulo the integer m

Numbers

$\lfloor x \rfloor$	'floor', or integer part, of x: the largest integer k such that $k \leq x$
$\lceil x \rceil$	'ceiling' of x: the smallest integer k such that $k \geq x$

$a_n \uparrow$	$a_1 \leq a_2 \leq \ldots$
$a_n \downarrow a$	$a_1 \geq a_2 \geq \ldots$ and $a_n \to a$
$a_n = O(b_n)$	a_n/b_n is bounded
$a_n = o(b_n)$	$a_n/b_n \to 0$ as $n \to \infty$
$|x|$	Euclidean norm, $\sqrt{x_1^2 + \cdots + x_n^2}$, of $x = (x_1, \ldots, x_n)^\top$
$\log$	base-2 logarithm: $x = \log y$ iff $y = 2^x$
$\log_+$	$\log_+ x = \log x$, if $x > 0$, or 0, if $x = 0$
$\ln$	natural logarithm: $x = \ln y$ iff $y = e^x$
$\log_a$	base-a logarithm: $x = \log_a y$ iff $y = a^x$
$a \equiv b \bmod m$	$a = b + km$ for some integer k

Vectors, strings, matrices

$a^\top$	transpose of vector a
$M^\top$	transpose of matrix M
$(a_1, \ldots, a_n)$	row n-vector
$(a_1, \ldots, a_n)^\top$	column n-vector
$a_1 a_2 \ldots a_n$	n-string
A^n	n-fold Cartesian product of set A, **or** n^{th} power of matrix A
A^*	set of all strings over alphabet A
$|c|$	length of string c (Chapter 1 only)
$\rho(x,y)$	Hamming distance, $\sum_1^n \mathbf{1}\{x_t \neq y_t\}$, between n-strings $x = x_1 x_2 \ldots x_n$ and $y = y_1 y_2 \ldots y_n$
$B(x,r)$	closed Hamming ball, $\{y : \rho(x,y) \leq r\}$, of radius r about x
$V_q(n,r)$	number of elements of $B(x,r)$, in a q-letter alphabet
$w(x)$	weight of $x \in F^n$
(x,y)	scalar product $\sum_1^n x_i y_i$
$x \wedge y$	wedge product $(x_1 y_1, \ldots, x_n y_n)$
I_k	$k \times k$ identity matrix

Sources, messages, channels, codes

$\mathbf{U}$	source
$u^{(n)}$	source n-word $(u_1, \ldots, u_n)^\top$ or $u_1 u_2 \ldots u_n$
$U^{(n)}$	random source n-word $(U_1, \ldots, U_n)^\top$ or $U_1 U_2 \ldots U_n$
$\mathcal{U}_n$	set of possible values of $U^{(n)}$
m_n	number of elements of $\mathcal{U}_n$
$\mathcal{C}_n$	set of n-strings allowed as input to channel
γ_n	code or coding $\mathcal{U}_n \to \mathcal{C}_n$
Γ_n	random code $\mathcal{U}_n \to \mathcal{C}_n$
$\mathcal{X}_n$	set of codewords: image of $\mathcal{U}_n$ under γ_n
$x^{(n)}$	transmitted codeword $x_1 x_2 \ldots x_n = \gamma_n(u^{(n)})$
$X^{(n)}$	random transmitted codeword $X_1 X_2 \ldots X_n$, $= \gamma_n(U^{(n)})$ or $\Gamma_n(U^{(n)})$
$\mathcal{Y}_n$	set of possible received n-words from channel
$y^{(n)}$	received word from channel: $(y_1, \ldots, y_n)^\top$ or $y_1 y_2 \ldots y_n$

$Y^{(n)}$	random received word from channel: $(Y_1, \ldots, Y_n)^\top$ or $Y_1 Y_2 \ldots Y_n$
$\hat{U}^{(n)}$	estimate of $U^{(n)}$
$\beta(\gamma_n)$	error probability $P(\hat{U}^{(n)} \neq U^{(n)})$ under coding γ_n, for equidistributed source-words and maximum-likelihood decoding
C	channel capacity (Chapters 3–4), code (Chapters 5–6)
$[n, m]$	code parameters: length n, size m
$[n, m, d]$	code parameters: length n, size m, minimum distance d
(n, r)	linear code parameters: length n, rank r
(n, r, d)	linear code parameters: length n, rank r, minimum distance d
$\rho(C)$	information rate of code C
$d(C)$	minimum distance of code C
$C^\perp$	linear code dual to linear code C
C^+	extension of linear code C
C^-	truncation of code C
C'	shortening of code C
$C^{(m)}$	m-fold repetition of code C
$C_1	C_2$	bar product of linear codes C_1 and C_2, where $C_1 \supseteq C_2$
$A_q(n, d)$	maximum size of code of length n with minimum distance d over a q-symbol alphabet
$A(C, x, y)$	weight enumerator of linear code C
$RM(d, r)$	Reed-Muller code of order r, length $n = 2^d$

Entropy, information

$h(U)$, $h(p)$	entropy of distribution $p = (p_.)$ of random variable U
$H = H(\mathbf{U})$	information rate of source $\mathbf{U}$
$h(X, Y)$	joint entropy of random variables or vectors X, Y
$h(Y|X)$	conditional entropy of Y given X
$I(X \wedge Y)$	mutual information between X and Y
$h_c(X)$	continuous entropy of continuous random variable X
$H_q(x)$	entropy function: $= 0$ if $x = 0$, and

$$x \log_q(q - 1) - x \log_q x - (1 - x) \log_q(1 - x)$$

if $0 < x < 1 - q^{-1}$

Probability

$P(A)$	probability of event A
$P(U = u)$	probability of $\{\omega : U(\omega) = u\}$, the event that $U = u$
p_X	probability mass function, $p_X(x) = P(X = x)$, of random variable X
$E(S)$	expectation or mean, $\sum_s s P(S = s)$, of random variable S
$\mathrm{var}(X)$	variance of random variable X
$P(A|B)$	conditional probability of event A given event B
$f_{Y|X}(\cdot|x)$	conditional density for Y given $X = x$
$E(Y|X = x)$	conditional expectation of Y given $X = x$

$\xrightarrow{\text{a.s.}}$	almost-sure convergence
$\xrightarrow{P}$	convergence in probability
$\xrightarrow{L^1}$	convergence in L^1
$\xrightarrow{L^2}$	convergence in L^2
$\xrightarrow{d}$	convergence in distribution
$N(0,1)$	standard Gaussian or normal distribution
$N(\mu, \sigma^2)$	Gaussian distribution with mean μ and variance σ^2
ϕ	density of $N(0,1)$
Φ	distribution function of $N(0,1)$
p_{jk}	transition probability of Markov chain, **or** element of channel matrix
$p_{jk}^{(r)}$	r-step transition probability of Markov chain

Algebraic objects

$\langle x \rangle$	principal ideal with generator x
$R[X]$	ring of polynomials in X over a ring R
∂f	degree of polynomial f
$x + W,\; W + x$	coset, with representative x, of subgroup W in an additive group
$R/\langle x \rangle$	quotient ring
$y \equiv z \bmod x$	$y = z + t\langle x \rangle$ in ring R
$X \oplus Y$	direct sum of subspaces X, Y
V^*	vector space dual to vector space V
U°	annihilator of subspace U
$\mathrm{GF}(q)$	finite field, of prime-power order q

INDEX

Proper names appear in the index only where not attached to a referenced book or paper.

Printed in the United States
By Bookmasters